Modern Farm Power

Modern Farm Power

Third Edition

William J. Promersberger

Professor Emeritus
Agricultural Engineering Department
North Dakota State University

Donald W. Priebe

Chairman, Agricultural Education Department
North Dakota State University

Frank E. Bishop

Former Instructor, Vocational Agriculture
Harvey, North Dakota

Reston Publishing Company
A Prentice-Hall Company
Reston, Virginia

Library of Congress Cataloging in Publication Data

Promersberger, William J
  Modern farm power.
  Includes bibliographical references and index.
  1. Farm engines. 2. Farm tractors. 3. Farm mech-
anization. I. Bishop, Frank E., joint author.
II. Priebe, Donald W., joint author. III. Title.
S675.P687 1979   681′.763   78-31917
ISBN 0-8359-4560-X

10 9 8 7 5 4 3 2 1

Printed in the United States of America

About This Book

There have been a great number of very important changes in technology. Many of these significant developments have taken place in the field of farm power and closely related areas. Such progress and change pose a greater need for a thorough understanding of internal-combustion engines, and allied equipment and accessories in the field of farm power. This need has led to the revision of this book and the preparation of this third edition.

Included in the third edition are a new chapter on tractor types and trends, and a discussion on Nebraska Tractor Tests and their significance. Solution to problems are shown in metric as well as English units.

Emphasis has been placed on many new developments including the alternator and newer forms of ignition systems. A changing emphasis in air cleaner systems and the increasing importance of hydraulic systems are reflected.

The study questions at the end of the chapters have been revised and expanded. The shop exercises have been revised and some new exercises have been added.

This book has been prepared as a helpful text or reference for classes in farm power. It will be equally useful to vocational agriculture instructors, farmers, mechanics, and others who operate, maintain, or service power units.

We present a brief history of engines, followed by an explanation of engine-operating principles, in the first two chapters. Next, the parts of an engine are identified and their functions detailed. An exposition of the fundamentals of machines, the fuels that give them power, and the principles of combustion that activate these fuels will add to your knowledge. We have discussed the important subjects of valves and fuel systems at some length.

Other vital parts of farm machinery, the air cleaner on the engine and the governor which regulates engine speeds, are explained. Means by which the fuel charge is ignited and the assorted electrical accessories are considered next. The chapter on diesel engines treats this important area. Engine cooling and lubricating systems are also important matters in the overall knowledge of maintenance and repair.

The drive and steering mechanisms on tractors require detailed explanations. The subjects of power-takeoff shafts and hydraulic systems which make the tractor such a versatile farm machine are given strong emphasis.

A chapter on safe practices in using mechanized equipment is based on the recommendations of the National Safety

Council. A discussion on noise levels is also included.

Advice on selecting tractors and machines, setting up a farm shop for their maintenance, and housing and storing them concludes this text.

Students need not have a technical background to use this book, although courses in basic sciences and mathematics will be useful. We trust that the third edition of *Modern Farm Power* will help the owners and operators of farm machines to obtain the power, efficiency, and service that the manufacturers have built into them.

We should like to acknowledge the help received from Henry L. Kucera and Roger H. Cossette, and from the companies and organizations that provided us with material and illustrations.

WILLIAM J. PROMERSBERGER
DONALD W. PRIEBE
FRANK E. BISHOP

Table of Contents

1 A Brief History of Engines

The study of mechanical power covers a broad area of learning. Whether you are beginning or advancing in the field, this study offers interesting challenges to those who will put forth the time and effort to acquire a working knowledge of internal-combustion engines. A basic understanding of the fundamentals of engines is absolutely necessary if we are to keep pace with farm power advancements.

Figure 1-1 shows an early engine and Figure 1-2 a modern internal-combustion engine. If you compare these, you will find that many changes have been made. However, the basic principles of operation have remained the same for many years.

FIRST EXPERIMENTS

As early as 100 B.C., Hero, an Egyptian, built a simple engine that made use of steam power to spin a hollow metal ball. This device was only a toy. In the twelfth century, experimenters who hoped to be able to get power out of exploding materials tried to use gunpowder as a fuel, with impractical and often disastrous results.

Denis Papin, a Frenchman, developed the first gas engine in 1690. However, this engine could not produce useful power. In 1705, Thomas Newcomen, an Englishman, invented the first "atmospheric engine." The Newcomen engine used steam and atmospheric pressure to pump water and is credited with being the first successful engine.

James Watt, a Scottish instrument maker, improved the Newcomen engine, incorporating a throttle, a flywheel, a governor, and a steam safety valve. Watt developed many devices used on present-day engines. Figure 1-3 illustrates the operating principles of the steam engine.

EARLY GAS ENGINES

Etienne Lenoir, a Frenchman, invented the first gas engine in 1860. Early experimenters used mixtures of coal gas, turpentine vapor, and air as combustible fuel. In 1862, another Frenchman, Beau de Rochas, patented an engine in which gaseous fuel was compressed by a piston. It was not until about 1876 that this principle was applied to a gas engine, when Nikolaus Otto, a German scientist, invented the four-cycle gas engine. Otto's engine used less gas and developed more power, which made this engine practical for commercial use. This, the first internal-combustion engine that worked well, is often referred to as the *Otto cycle engine*. One of his later engines (1884) is illustrated in Figure 1-1.

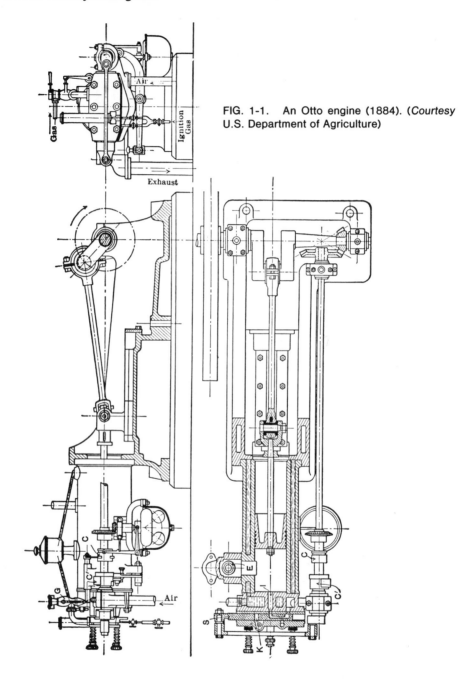

FIG. 1-1.   An Otto engine (1884). (*Courtesy* U.S. Department of Agriculture)

Taking advantage of ideas and patents developed by Otto, Sir Dugald Clerk invented the two-cycle engine in 1881.

In 1890, Rudolf Diesel, a German engineer, adapted ideas of the early combustion engines to an engine that used the heat of high compression to ignite the fuel charge. In 1892 he obtained a patent and in 1897 completed the building of his first engine, which is now known as the diesel engine. However, the first diesel engine to be used on a farm tractor did not appear until about 1931.

Liquid petroleum gas tractors, to be referred to later in this book as LPG tractors, came into use in farming and industry in 1941.

DISCOVERY OF OIL AND ITS RELATION TO ENGINES

The moving parts of any engine require lubrication to maintain their performance and extend their service life. However, early engines ran very slowly, making the need for lubrication less of a problem than in the faster-running, more powerful engines developed later.

For the latter, a high-quality lubricant was needed to reduce engine friction, provide cooling, and cushion the force of exploding gases. Whale oil had furnished man with a lubricant prior to the beginning of the petroleum industry.

FIG. 1-2. A modern internal-combustion engine. (*Courtesy* U.S. Department of Agriculture)

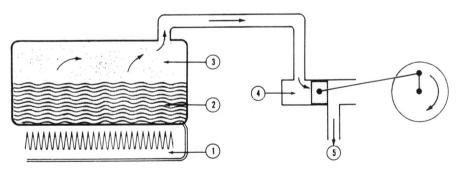

FIG. 1-3. Operating principles of the steam engine: source of heat (1) causes the water (2) to produce steam (3) which passes through a tube to the piston chamber (4). Steam pressure forces the piston outward to turn a flywheel, thereby changing reciprocating motion to rotary motion. Through momentum, the weighted flywheel returns the piston to its original position after the expanded steam pressure has escaped through the exhaust tube (5). (Drawing by Roger Cossette)

One whale could provide as many as 200 barrels of oil.

Petroleum deposits which had appeared on the earth's surface through natural forces were man's first encounters with another kind of oil. The first known oil well was dug in Pennsylvania in 1859. People of the time used this crude oil in lamps. Explosions of these lamps showed the presence of volatile gas and led to the problem of separating this combustible gas from the less combustible oil. This was done by a process called distillation, which resulted in an engine fuel as well as an oil for lubrication.

Oil reduced friction in early engines, and helped provide a seal between the piston and the cylinder, making the engines more efficient and powerful. Solving the friction and heat problems encouraged the development of power units with higher horsepower.

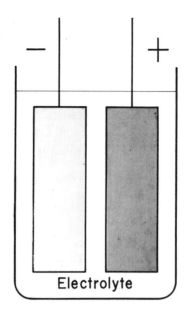

FIG. 1-4.   The Volta cell. (Drawing by Roger Cossette)

ELECTRICITY AND ITS RELATIONSHIP TO ENGINES

Early inventors of gasoline engines were constantly searching for greater efficiency and a satisfactory method of igniting the fuel charge in the combustion chamber. Open flames, glow plugs, and red-hot wires were used to fire fuel charges. Today we obtain far better results through the use of electricity.

The first device for producing electricity was invented by a German physicist, Otto von Guericke, in 1672. Volta, an Italian physicist, developed the first storage battery (Figure 1-4).

This battery principle was improved upon and enlarged until a battery of sufficient size was built to be practical for engine use. Battery voltage must be intensified to cause it to jump an air gap between two electrodes, as it does on a spark plug. A coil is used to step up battery voltage to a point high enough to jump across a spark plug gap. These electrical principles were adapted to engines by placing the electrodes of the spark plug in the engine combustion chamber, in order for the spark to ignite or fire the compressed air-fuel mixture.

Producing electrical energy and storing it in a battery were important developments in the improvement of the internal-combustion engine.

EFFECT OF ENGINES ON THE MECHANIZATION OF AGRICULTURE

When engines were in the early developmental stage in the beginning of the nineteenth century, the sources of power for farming were mainly man and animals. Animals were used for field work and some farmstead chores. During this time most of the people in our country were engaged in farming. This was necessary to provide enough food for themselves and the rest of the population.

The development of steam engines in the latter part of the nineteenth century made possible the mechanization of some tillage and harvesting operations. However, steam engines were heavy and awkward to handle in the field. The more versatile and lighter gasoline and diesel engines were developed for farm use in the early part of the present century. They have had a tremendous influence on the mechanization of our agriculture. Today, with modern equipment, less than five percent of our population are engaged in farming and are producing more than enough food for the entire population of our country.

Mechanization has helped make it possible to produce food on the farm with a great reduction in the amount of labor required. In about 1800, when hand labor was used for most farm operations, about 55 man hours were required to produce an acre of wheat. When horses and steam engines were used in about 1875 the labor required to produce an acre of wheat was reduced to about 15 man hours. In 1925, when gasoline tractors had replaced many horses, the labor required to produce an acre of wheat was about 5 man hours. At the present time with modern equipment and tractors, wheat can be produced with a labor requirement of only about 1 or 2 man hours per acre.

It is the gasoline (spark ignition) and diesel engines and tractors that have had such a great influence on the mechanization of our agriculture, that we want to discuss in these chapters.

Questions

1. Name the early experimenters who are credited with inventing the following:

   (a) first engine

   (b) first gas engine

   (c) four-cycle engine

   (d) two-cycle engine

   (e) diesel engine

2. Why was the discovery of oil important to the development of the engine?

3. Who invented the first device for producing electricity? For storing electricity?

4. Why was the discovery of electricity a step forward in engine development?

5. What effect has the development of modern engines and tractors had on agriculture?

6. Explain how the development of internal combustion engines influenced the labor requirements for crop production.

References

Collier's Encyclopedia with Bibliography and Index, Crowell Collier Educational Corporation, New York, New York, 1975.

Funk and Wagnalls New Encyclopedia, New York, New York, 1975.

Promersberger, W. J., *More Time To Live,* Sixteenth Annual Faculty Lectureship, North Dakota State University, Fargo, North Dakota, February 15, 1972.

2 Engine Operating Principles

Internal-combustion engines generate power by utilizing the force created by burning a mixture of fuel and air. This force is confined or trapped in a combustion chamber. The expanding gases force the piston downward in the cylinder. Because the piston is connected to a crankshaft by a connecting rod, this downward motion is changed to a rotating motion by mechanical linkage.

The momentum of the moving parts, aided by a weighted flywheel, returns the piston to its former position to receive the explosive power of the fuel-air mixture. Gasoline, diesel, and liquefied petroleum gas (LPG) engines use many of the same operating principles. Where a difference in principle occurs, it will be specially noted.

FOUR-STROKE CYCLE ENGINE

The four-stroke cycle engine is the most common type found on farms. An engine stroke is commonly thought of as the movement of the piston from Top Dead Center (TDC) to Bottom Dead Center (BDC). TDC is reached when the piston is at the end of the inward stroke, BDC is at the end of the outward stroke. A stroke can also be defined as one-half of a revolution or 180° of crankshaft travel. An engine cycle is the complete set of movements necessary to generate engine power (Figure 2-1). The four strokes in a cycle are: *intake, compression, power,* and *exhaust.* How does each stroke fit into the engine cycle? What happens on each stroke? These questions must be answered and the answers understood before we can develop a knowledge of the principles of operation of an internal-combustion engine.

Intake

With the intake valve open, the piston moves outward in the cylinder (toward the crankshaft), causing a partial vacuum, which draws a mixture of fuel and air into the cylinder (in a diesel, only air is taken in).

Compression

As the piston reaches BDC and begins its inward motion (toward the cylinder head), both valves are closed and the piston compresses the mixture between the piston and the cylinder head (only air is compressed in the diesel).

Power

As the piston nears TDC, an electric spark ignites the compressed mixture, pushing the piston downward with great

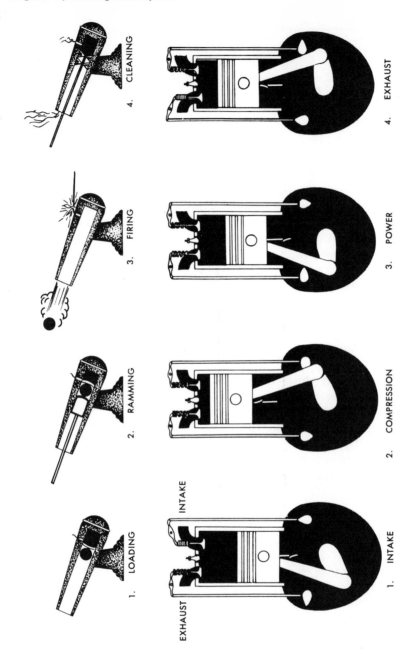

FIG. 2-1.   Operating principles of a four-stroke cycle engine. The spark plug ignites the fuel charge. (*Courtesy American Society of Agricultural Engineers*)

force (in the diesel engine, fuel is injected into highly compressed air, causing self-ignition).

Exhaust

The fourth and last stroke of the cycle occurs when the piston begins its upward movement. The exhaust valve opens, allowing the piston to force out the burned gases, thus clearing the cylinder for the start of another cycle of 720° of crankshaft travel. Since this cycle

of four strokes—intake, compression, power, and exhaust—is repeated time after time, the engine is properly called a four-stroke cycle engine.

TWO-STROKE CYCLE ENGINE

The intake, compression, power, and exhaust strokes of the two-cycle engine occur in 360° of crankshaft travel. Ports or openings are used instead of valves in this engine (Figure 2-2). The

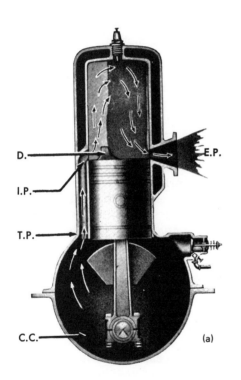 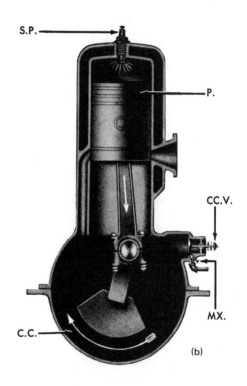

FIG. 2-2. Operating principles of the two-cycle engine. (a) End of the downstroke when the exhaust gases escape through the exhaust port (E.P.), while a new fuel charge is drawn or forced into the cylinder through the intake port (I.P.). The piston deflector (D.) directs the gases in the cylinder. (b) The piston in the full-up or TDC position, where combustion takes place and the power stroke begins. The crankcase (C.C.) is the precombustion chamber where the air-fuel mixture (MX.) is taken in through the crankcase valve (CC.V.) as the piston moves toward TDC, creating a partial vacuum in the crankcase. As the piston moves toward BDC, compression takes place in the crankcase, forcing the mixture through the intake port. (*Courtesy Standard Oil Company of Indiana*)

power stroke takes place in the same manner as in the four-stroke cycle engine, but near the end of this stroke the exhaust port is uncovered by the piston and the exhaust gases escape. Immediately following this, the intake port is uncovered and the new fuel charge is admitted from the crankcase, where it has been slightly compressed. The compression stroke follows and the engine fires near TDC. The cycle is then repeated. The two-cycle engine utilizes an airtight crankcase for partially compressing the fuel-air mixture. The pressure differential between the crankcase and the atmosphere allows the air to rush into the cylinder from the crankcase and force out the burned exhaust gases. A baffle on the pistonhead directs the flow of air into and out of the cylinder. An electrical spark ignites the compressed mixture and a power stroke occurs on each revolution of the engine.

An air-fuel mixing valve and an exhaust port located at the bottom of the cylinder are means of identifying the two-cycle engine. There are no intake or exhaust valves. (Some two-cycle diesel engines have exhaust valves. These engines will be discussed in another chapter.)

Since the crankcase is used as a pressure chamber for the incoming fuel charge, lubrication of the engine is accomplished by mixing oil with the fuel. Approximately one-half pint of oil per gallon of fuel is required for proper lubrication. Thorough mixing of the oil and fuel is important before placing this mixture in the engine fuel tank.

Because of back pressures created in long exhaust pipes, this type of installation should be avoided on a two-cycle engine. This engine is usually less efficient than the four-cycle type. Keeping the crankcase airtight and the exhaust port clean will aid in proper engine starting and operation.

LPG engines are carburetor-type engines. The principles of operation are the same except for carburetion and fuel handling. This will be discussed in Chapter 6.

ENGINE EFFICIENCY

The mechanical efficiency of an engine is its ability to change fuel to usable horsepower. Whenever the heat or energy of a fuel is converted or changed from one form to another, a considerable part of it is lost. Of each four gallons of fuel added to a carburetor-type engine, about three gallons are lost. This gives the engine an efficiency of approximately 25 percent. Diesel engines have a higher efficiency. Drawbar horsepower and power take-off (PTO) horsepower are common measures of a tractor's ability to do work. Drawbar horsepower is about 15 percent less than PTO horsepower because of losses in the transmission, differential, and drive wheels.

HOW IS POWER LOST IN AN ENGINE?

Power losses in the engine are through the exhaust system, the cooling system, and engine friction, and in transmission and traction losses. Approximate losses in each are shown in Figure 2-3. The heat energy lost in the exhaust gases (31 percent) is added to the heat lost in the cooling system (40 percent) and that lost in engine friction (5.5 percent). This amounts to a total loss of 76.5 percent of the starting fuel. The remaining 23.5 percent is available for useful work at the power take-off (PTO)

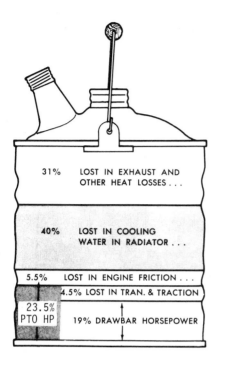

FIG. 2-3. Approximate heat-energy losses in a tractor engine. (Drawing by Roger Cossette)

shaft; however, another 4.5 percent is lost between the PTO shaft and the tractor drawbar.

SUMMARY

Most farm engines are of the four-stroke cycle type, although some of the smaller engines are of the two-stroke cycle type. The two-cycle diesel engine is used in some farm tractors. To get the most from farm engines, proper servicing and maintenance are most important. Engines that are properly maintained and are operated at proper temperatures will give greatest efficiency, long engine life, and low operating costs.

The four-stroke cycle is made up of the following movements. On the *intake stroke,* the movement of the piston from TDC to BDC results in the taking in of an air-fuel mixture. *Compression* follows intake and occurs when both valves are closed. The air-fuel mixture is compressed into the combustion chamber as the piston moves from BDC to TDC. *Power* is obtained from the expanding gases after the spark plug has ignited the compressed mixture. The force of the expanding gases on the piston head is transmitted through the connecting rod to the rotating engine crankshaft as the piston moves from TDC to BDC. *Exhaust* occurs when the exhaust valve opens and the piston moves from BDC to TDC, pushing out the burned gases.

Four-stroke cycle engines provide more time for this series of events to take place than do two-stroke cycle engines. The completeness of each stroke during the 720° of crankshaft travel results in greater efficiency than that obtained when intake, compression, power, and exhaust must take place in 360° of crankshaft travel, which happens in the two-stroke cycle engines.

Spark plugs are used to ignite the air-fuel mixture in the gasoline engine. The temperature of the air is raised to the kindling temperature of the fuel by compression in the diesel engine. This temperature is approximately 1100° F.

Much of the fuel consumed by an engine is not used to propel the machine or to pull a load. Of each gallon of fuel burned, 31 percent is lost through the exhaust system, 40 percent through the cooling system, 5.5 percent through engine friction, and 4.5 percent through power transmission to the drive wheels. In the gasoline engine, this leaves approximately 19 percent for load purposes. A diesel engine has a slightly higher efficiency. This superior efficiency is basically due to its higher compression ratio and also to the fact that it burns a fuel that has a greater heat content per gallon.

Shop Project:
Engine Study

The objective of this project is to study the two-stroke cycle and four-stroke cycle engines in order to determine differences in construction and principles of operation.

A. Use small one-cylinder engines for this study, one four-stroke cycle engine and one two-stroke cycle engine.

B. Study the construction of the two engines.

1. Fuel system

(a) *Four-cycle* engine—The fuel flows from the fuel tank to the carburetor (or air-mixing valve) to the intake valve to the combustion chamber. The exhaust gases escape through the exhaust valve. The cylinder head must be removed to see the intake and exhaust valves. Consult with your instructor before removing the cylinder head.

(b) *Two-cycle* engine—The fuel flows from the fuel tank to the carburetor (or air-mixing valve) to the engine crankcase to a passageway to the intake port and to the combustion chamber. The cylinder head (or block) must be removed to see the passageway and inlet port. The exhaust port is visible from the outside of the engine and is located at the lower end of the cylinder wall.

2. Valves

(a) Turn the crankshaft of the four-cycle engine and notice the operation of the valves.

The intake valve controls the passage between the carburetor and the combustion chamber.

The exhaust valve controls the passage between the combustion chamber and the exhaust manifold.

3. Ports

    (a)   Examine the ports near the bottom of the cylinder on the *two-cycle* engine. (The *two-cycle* engine does not have valves.)

         The intake port leads from the crankcase to the combustion chamber.

         The exhaust port leads directly outside through a short exhaust pipe.

         Also notice that, as the piston moves downward (out of the cylinder), the exhaust port is uncovered first. This gives most of the exhaust gases a chance to escape before the intake port is uncovered.

4. Reassemble the engines. Use new gaskets wherever necessary.

Questions

1. What is a four-stroke cycle engine?

2. What is a two-stroke cycle engine?

3. What actually takes place on each of the following strokes?

    (a)   intake

    (b)   compression

    (c)   power

    (d)   exhaust

4. How do two-cycle and four-cycle engines differ in construction and in operation?

5. What are the advantages and disadvantages of the two-cycle engine?

6. What is the meaning of engine efficiency?

7. Where do power losses occur in engines?

References

Care and Operation of Small Engines, Volume 1, American Association for Vocational Instruction Materials, Athens, Georgia, 1975.

Fundamentals of Service, Engines, 3rd Edition, Deere & Company, Moline, Illinois, 1977.

Roth, Alfred C. et al, *Small Gas Engines,* The Goodheart-Wilcox Company, Inc., South Holland, Illinois, 1975.

3 Identification and Function of Engine Parts

Internal-combustion engines have many parts. In order that we may identify and understand the function of each, we will divide these parts into four basic groups:

1. stationary parts
2. rotating parts
3. reciprocating parts
4. engine accessories

Let us examine the various parts and determine the function of each. Figure 3-1, A, B, and C, shows the stationary parts.

STATIONARY ENGINE PARTS

The *block* is a group or bank of cylinders cast in one unit, usually made of cast iron, containing water jackets that surround the cylinders to aid in cooling.

The *cylinders* are tubes in the engine block that have very smooth interior finishes and serve as guides for the pistons as they move up and down.

The *crankcase,* or lower part of the block, confines the lubricating oil near the engine's moving parts and supports the crankshaft and camshaft bearings.

The *oil pan,* or oil reservoir, aids the crankcase in confining the oil near the engine moving parts. The pan is made of pressed steel and bolts onto the bottom of the crankcase.

The *cylinder head* is a cap that attaches to the top of the engine block and covers the upper cylinder openings, thereby forming a combustion chamber in the engine.

ROTATING ENGINE PARTS

Figure 3-2, A and B, shows the rotating parts of the engine. These parts travel with a circular movement, one of the two basic types of motion found in the engine.

The *crankshaft* changes reciprocating action of the pistons to rotating motion. This action makes a turning force, or torque, available to the driving members of the tractor. The crankshaft is usually made of cast or forged steel. It is attached to the crankcase through main bearings. Throws, or offsets, are the areas between the journals and the crankpins. Pistons are attached to the crankshaft through connecting rods. The connecting rods are attached to the crankshaft at the crankpins by connecting rod bearings. Counterweights are often attached to the crankshaft to make possible precise balancing of this unit.

A *flywheel* is attached to the crankshaft and serves four useful purposes: (1) it smooths out engine power impulses, (2) it can be connected to the starter for starting purposes, (3) it pro-

FIG. 3-1.   Stationary engine parts. (A) Cylinder block. (B) Oil sump. (C) Cylinder head. (*Courtesy* Allis-Chalmers Manufacturing Company)

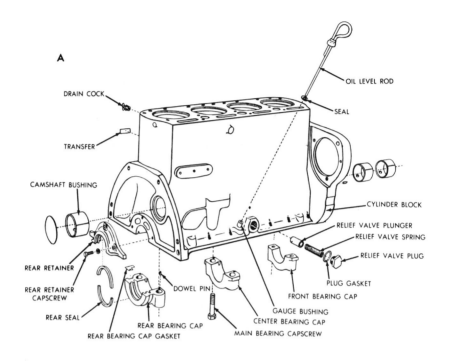

A

OIL LEVEL ROD

DRAIN COCK

SEAL

TRANSFER

CAMSHAFT BUSHING

CYLINDER BLOCK

RELIEF VALVE PLUNGER
RELIEF VALVE SPRING
RELIEF VALVE PLUG

REAR RETAINER

PLUG GASKET

REAR RETAINER CAPSCREW

DOWEL PIN

FRONT BEARING CAP

REAR SEAL

GAUGE BUSHING
CENTER BEARING CAP

REAR BEARING CAP

REAR BEARING CAP GASKET

MAIN BEARING CAPSCREW

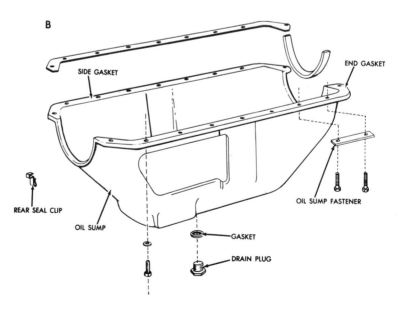

B

SIDE GASKET

END GASKET

OIL SUMP FASTENER

REAR SEAL CLIP

OIL SUMP

GASKET

DRAIN PLUG

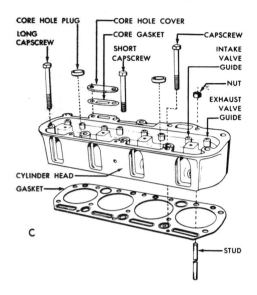

FIG. 3-1. continued

vides a place to mount the clutch, and (4) it provides an extra turning force, inertia, when a sudden load is applied to the engine. Timing marks are often located on the flywheel for the purpose of timing the ignition system to the engine.

A *camshaft* is a lobed shaft providing eccentric action for opening the valves. A cam is defined as a piece of metal that does not turn about its center. Figure 3-3 shows the action of a cam.

The camshaft is driven from the crankshaft by a timing gear, or through a timing chain. Small dots or other marks are found on these gears to permit proper assembly and timing of the valve mechanism with the pistons of the engine.

Since the two-cycle engine employs ports rather than valves, no camshaft will be found in this engine.

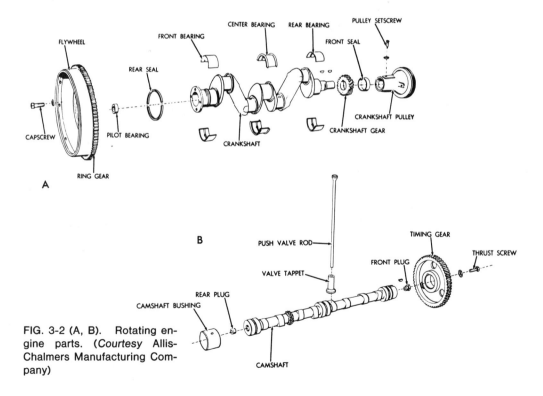

FIG. 3-2 (A, B). Rotating engine parts. (*Courtesy* Allis-Chalmers Manufacturing Company)

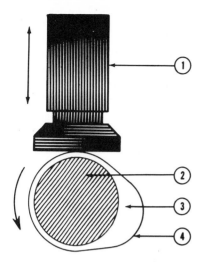

FIG. 3-3.   Valve lifter (1), cam (3), and cam-shaft-action (2). As the lobed surface (4) of the cam rotates and comes in contact with the valve lifter, a reciprocating action takes place. (Drawing by Roger Cossette)

RECIPROCATING ENGINE PARTS

Reciprocating parts of the engine are shown in Figure 3-4. These parts stop, start, or change direction frequently and are one of the main causes of engine vibration.

Pistons are made of aluminum alloy or cast iron. The pistons and associated parts are shown in Figure 3-4. The force of the exploding gases is received by the piston head and transmitted to the piston pin, connecting rod, and crankshaft.

Rings are fitted to the piston to seal in combustion and compression pressures, as well as to prevent oil from the crankcase from entering the combustion chamber.

Valves open and close the ports in the combustion chamber. Since there are two openings or ports for each cylinder, there must be two valves. An intake valve allows the fuel-air mixture (in the diesel engines, air only) to enter the combustion chamber when the valve is open. Exhaust valves open to allow burned gases to pass from the combustion chamber into the exhaust system. Both valves close on the compression and power stroke to confine the gases to the combustion chamber.

Connecting rods connect the pistons to the crankshaft. The reciprocating action of the piston is changed to rotating

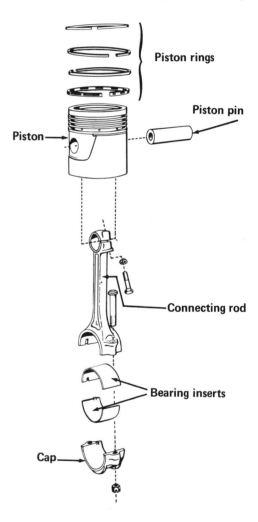

FIG. 3-4.   Reciprocating engine parts. (*Courtesy* Allis-Chalmers Manufacturing Company)

action at the crankshaft through the bearing surfaces on either end of the connecting rod.

Valve seats in the cylinder head (I-head engines) or engine block (L-head engines) contain sealing surfaces that contact the valve face to provide the sealing-off effect. Figure 3-5 shows the actuating mechanism of the overhead-valve rocker-arm system.

ENGINE ACCESSORIES

Four separate systems are necessary for proper operation of the internal-combustion engine. They are electrical, fuel, lubricating, and cooling. Accessory parts of the engine are components of these systems. Each will be discussed in more detail in other chapters. They are briefly mentioned here.

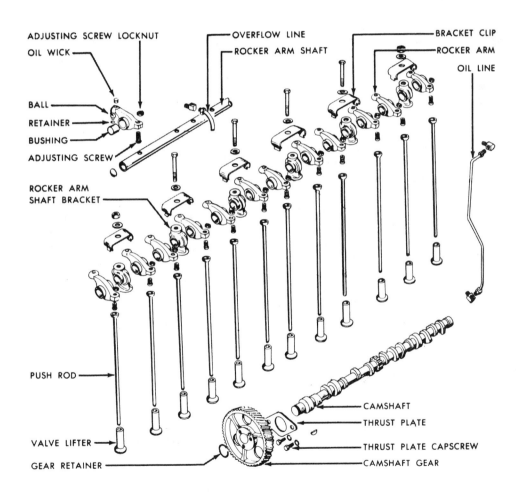

FIG. 3-5. The overhead-valve mechanism is actuated by the camshaft through a valve-lifter, pushrod, rocker-arm arrangement. (*Courtesy* Allis-Chalmers Manufacturing Company)

The Electrical System

Electrical current is supplied by a *magneto* or a *storage battery*. Driven by the *timing gear*, the magneto produces high-tension voltage and distributes it to the *spark plugs* for ignition purposes. In the battery ignition system, the battery acts as the source of current. A *coil* serves as a spark intensifier and delivers high-tension voltage to a *distributor*, which in turn delivers the spark to the spark plugs at the proper time. Spark plugs provide an air gap across which this voltage must pass or jump to go to the ground in this system. This spark ignites the compressed air-fuel charge in the *combustion chamber*.

The Fuel System

Fuel supply tanks are located above the carburetor, with the fuel flowing through a *filter-sediment bowl* device to the carburetor by the force of gravity. Should the fuel tank be located below the carburetor, a *fuel pump* is necessary to supply fuel pressure to the carburetor. Diesel engines use a high-pressure pump to force fuel through injectors into the combustion chamber against the pressure of compression. Gasoline engines use a *carburetor* to mix the fuel and air into a combustible mixture, then deliver it to the engine through a unit called an *intake manifold*.

The Lubricating System

Engine lubrication is provided by an *oil pump* located in the *oil pan*. Pumps are of three types: vane, piston, and gear. Gear-type pumps are used in most engines because of their long life and trouble-free operation. *Pump inlets* are often found in a float that takes oil from the cleanest place in the oil pan. *Oil filters* are located between the oil pump and engine parts to remove abrasive particles and reduce engine wear.

The Cooling System

Cooling systems are of two types, *thermosiphon* and *pump*. In the thermosiphon system, water expands as it is heated, causing it to rise in the cooling system to the top of the radiator. Gravity acts on the water to cause downward movement in the radiator, where cooling takes place, then back to the engine water jackets to provide the cooling of engine parts. The *pump system* provides pressure for circulation, and water follows the same path as in the thermosiphon system.

CYLINDER ARRANGEMENTS

Internal-combustion engines used in tractors may have from two to eight cylinders. Twelve- and sixteen-cylinder engines are used in the field of industry. Regardless of the number of cylinders, engines are of two basic types, *in-line* and V-*type*. In-line engines have the cylinders set in an upright position, one directly behind the other. V-type engines have the cylinders arranged in banks of two, three, or four, set at an angle to each other to form a modified V.

CRANKSHAFT ARRANGEMENTS

The flow of power from the engine cylinders is not smooth. Various crankshaft arrangements make it possible to distribute the power being delivered to the crankshaft, as well as to balance the power and compression strokes. On a two-cylinder engine the crankpins are at

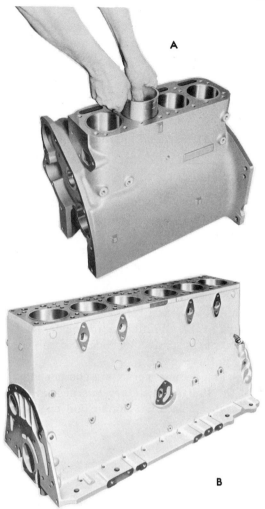

FIG. 3-6. (A) Four-cylinder inline engine block. (B) Six-cylinder inline engine block. (*Courtesy* Allis-Chalmers Manufacturing Company)

180°, or directly opposite each other. The four-cylinder engine has the crankpins at 180° arranged in pairs, as shown in Figure 3-7. This arrangement causes pistons 1 and 4 to travel up and down together and pistons 2 and 3 to move together. *Firing order* is the sequence of power strokes in the engine. Two firing orders are possible: 1-3-4-2 or 1-2-4-3.

Six-cylinder in-line engines have crankshafts supported by three or four main bearings. Crankpins are arranged in pairs, 120° apart. Cylinders are numbered from front to rear, and the firing order is usually 1-5-3-6-2-4 or 1-4-2-6-3-5. Six and eight cylinder crankshaft arrangements are also shown in Figure 3-7.

METALS USED IN ENGINE CONSTRUCTION

When an engineer or product designer selects the metals to be used in a modern engine, he selects them on the basis of their properties. These properties include the following: *Density,* which refers to the weight of the metals; *Corrosion Resistance,* the ability to resist rusting and other chemical action; *Hardness,* or the ability to resist penetration; *Tensile Strength,* which refers to the metals' resistance to being pulled apart; and *Toughness,* or the ability to withstand shock or heavy impact forces without breaking.

Metals are classified as pure metals or as alloys. A pure metal is a single chemical element which is not combined with any other element. Pure metals such as iron, copper, and aluminum are generally too soft, lack strength, and rank low in other desired properties. Alloys are combinations of two or more metals, or, a metal and nonmetallic elements. Steel is an alloy obtained by combining manufactured metals, or the combination of a metal and a nonmetal. Carbon steel is made by combining iron and carbon. Alloy steel can contain one or more of the following metals: nickel, chromium, manganese, molybdenum, tungsten, and vanadium. Aluminum is combined with copper, magnesium, zinc, manganese, and chromium in varying amounts to improve its physical properties. Copper is combined with tin, zinc, and nickel to form copper-base alloys.

Steel is used in engines where polished parts are needed or where hardened parts are required. Gears, shafts, and fasteners are examples of engine parts made of steel. Aluminum is a lightweight alloy that conducts heat rapidly and is used in pistons and some cylinder heads and engine blocks. Cast iron (iron + carbon) is a low-cost metal having excellent wearing properties and rigidity but lacking strength. Parts made of cast iron must be reinforced to withstand the forces applied to them. Engine blocks are usually made of cast iron. Copper conducts heat and electricity rapidly and is used where these properties are needed. Bushings and bearings are made of copper-based alloys.

Diesel engines must be made of heavier and stronger materials than gas-oline engines, since higher pressures and consequently greater forces are built up within the engine.

SUMMARY

The parts of an internal-combustion engine are divided into four groups: stationary, rotating, reciprocating, and engine accessory.

Stationary parts, as their name implies, do not move. They are the block, cylinders, cylinder head, and oil pan.

Rotating parts (crankshaft, flywheel, and camshaft) have a circular motion.

Reciprocating parts stop, start, or change direction frequently, causing much of the vibration in an engine. Pistons, rings, connecting rods, and valves are in this group.

Engine accessories are components of the four separate systems needed to operate an internal-combustion engine.

FIG. 3-7.   Crankshaft arrangements for four, six, and eight cylinder engines. (*Courtesy Deere and Company*)

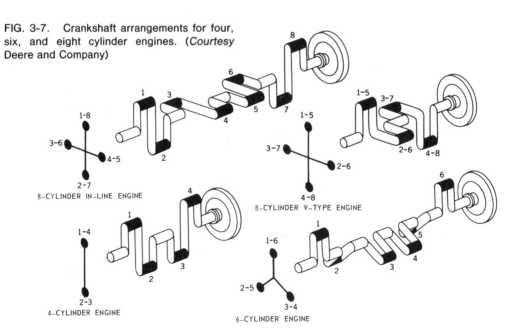

The systems are electrical, fuel, lubricating, and cooling.

There are two basic types of cylinder arrangement on tractor engines, in-line and V-type. Crankshaft and camshaft arrangements determine the firing order of the engine.

Steel, aluminum, cast iron, iron, and copper are the metals most commonly used in engines. The specific properties of each of these metals make them adaptable for specific purposes in the internal-combustion engine.

Shop Projects

A.   Examine crankshafts from four- and six-cylinder engines to determine which pistons move up and down together. Using the common firing order for six-cylinder engines (1-5-3-6-2-4) note how the power strokes are distributed on the crankshaft.

B.   For a four-cylinder four-cycle engine, start with No. 1 piston at TDC, power stroke, and list the stroke for each of the other three pistons. Do this by completing the following diagram. Assume a firing order of 1-3-4-2.

Degrees of Crank Travel	Cylinder Number			
	1	2	3	4
180°	Power			
360°	Exhaust			
540°	Intake			
720°	Compression			

Questions

1.   Into what four basic groups can we divide engine parts?

2.   Name the parts in each group and their functions.

3.   What four systems are necessary for proper engine operation?

4.   What is the purpose of each system?

5. Briefly describe the operation of each system.

6. How can crankshaft arrangement help smooth out engine operation?

7. What are the firing orders for four- and six-cylinder engines?

8. Name the metals used in engine construction, and two properties of each.

4 Fundamentals of Engines

At this point, we must add new terms to our vocabulary as well as develop an understanding of how certain basic scientific concepts are used in engine operation.

MEASUREMENTS

The units of measurement commonly used are known as the English units. These include the inch, foot, pound, cubic inch, gallon, degrees Fahrenheit etc. We are gradually converting to the metric system which includes the centimeter, meter, kilogram, liter, degrees centigrade etc. In this book we will use the English system, but in most cases we will also give the metric equivalent. When sample problems are shown they will be solved by the English system followed by the metric solution. This will provide an opportunity to become acquainted with the metric system.

A table of metric measures is shown in Appendix A.

Appendix B includes some conversion factors that will be useful in converting from one system to the other.

ATMOSPHERIC PRESSURE

Atmospheric pressure is an important concept. The atmosphere above and around us exerts a pressure on our bodies of 14.7 pounds per square inch at sea level. This pressure is present on all materials on the earth's surface. As we go from sea level to higher elevations, the atmospheric pressure decreases at a rate of approximately .4 pounds per 1000 feet up to an elevation of 3½ miles. "Pounds per square inch" is commonly referred to as *psi* and will be indicated in that way throughout this book.

If we know atmospheric pressure is 14.7 psi at sea level, we can appreciate, in the pressure per square foot, the tremendous force around us at all times. Because this force is present on all sides of a surface, its equalizing effect makes the force of atmospheric pressure go unnoticed in everyday life. As we progress in our study of farm power, we will see how very important this pressure is to the efficient operation of farm engines.

VACUUM

Vacuum is equally important in engine operation. It is a word used to define the absence of atmospheric pressure. We know that air will rush from areas of high pressure to areas of low pressure. The piston of an engine creates a partial vacuum when it moves downward, or toward BDC, in the cylinder; therefore, the internal-combustion

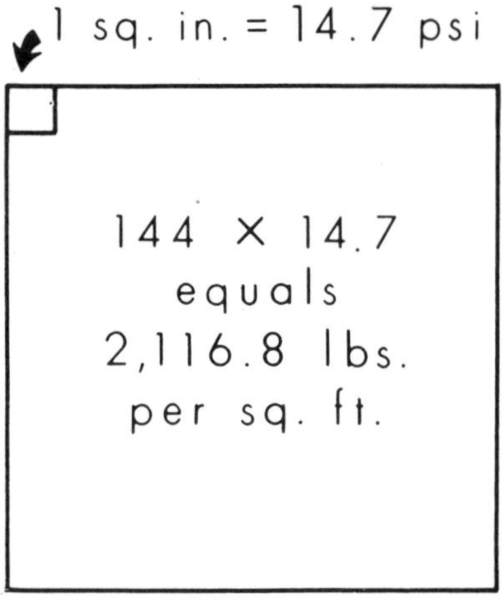

1 sq. in. = 14.7 psi

144 X 14.7
equals
2,116.8 lbs.
per sq. ft.

FIG. 4-1.   Atmospheric pressure per square inch. Atmospheric pressure per square foot. (Drawing by Roger Cossette)

engine is essentially a large vacuum pump. As the engine produces this vacuum, atmospheric pressure rushes through the carburetor air passage, mixing air with the engine fuel; this mixture then passes into each cylinder on the intake stroke.

Volumetric Efficiency

Just how efficiently an engine can produce vacuum and at the same time take in air at atmospheric pressure is termed the *volumetric efficiency* of the engine. This may be more properly defined as the percent of air-fuel mixture taken in compared with the amount that could be taken in under ideal conditions. Volumetric efficiency of an engine may vary from 50 percent at high speeds to 90 percent at low speeds. For example, a cylinder has a volume of 62 cubic inches. If this cylinder would fill completely on the intake stroke, .044 of an ounce of air would be taken in (since air weighs about 1.25 ounces per cubic foot, or ap-

proximately .0007 of an ounce per cubic inch). An engine running at half throttle would take in a smaller charge of the air-fuel mixture because the throttle is only partly open. As a result, it might take in only about .031 of an ounce of air. The volumetric efficiency of this engine at half throttle would then be about 70 percent (.031 is about 70 percent of .044).

WORK

Work is defined as the effect of a force in producing a change of position of an object against an opposite force. A force exerted through a distance provides a method of work measurement. A ten-pound weight lifted to a height of three feet represents 30 foot-pounds of work done. *Force times distance equals work.*

$$\text{Force (lbs.)} \times \text{Distance (ft.)} = \text{Work (ft. lbs.)}$$

POWER

Power is measured in terms of work accomplished in a given period of time. A large tractor can do more work in a given period of time than a small tractor, and, therefore, is said to be more powerful. The formula for determining power is:

$$\text{P (Power)} = \frac{\text{F(Force)} \times \text{D(Distance)}}{\text{T(Time)}}$$

HORSEPOWER

Horsepower is the most common measure of engine power. The term had its origin in the era when engines were competing with horses as a source of power. An engine's ability to do work was compared with that of a horse; if the engine could do the work of ten horses, the engine rating was ten horsepower. A horsepower is defined as the ability to do 33,000 foot-pounds of work in one minute. A horse that walks 100 feet in one minute while pulling with a force of 330 pounds is said to be developing one horsepower. The formula for horsepower is:

$$hp = \frac{foot\ pounds\ per\ minute}{33,000}$$

$$or \quad \frac{F \times D}{33,000 \times T}$$

In this formula, hp equals horsepower, D is distance in feet, F is weight or force in pounds, and T is time in minutes necessary to move F through distance D.

Problem: A farm implement requires a force of 2,640 lbs. to move it a distance of 500 ft. in one minute. What horsepower is required?

Solution: Substituting in the formula,

$$hp = \frac{F \times D}{33,000 \times T} = \frac{2640 \times 500}{33,000 \times 1} = 40\ hp$$

In the metric system power is expressed in kilowatts (kw). One horsepower is equivalent to 0.746 kilowatts. The answer in kilowatts to the example above is 40 × .746 = 29.84 kw.

BORE AND STROKE

The size of an engine cylinder is referred to in terms of *bore* and *stroke*.

The diameter of the cylinder is the bore. The distance the piston travels in the cylinder from TDC to BDC is called the stroke. When two measurements are given for a cylinder, the bore is given first. Thus a 4½″ × 5″ engine is one that has a 4½-inch bore and a 5-inch stroke.

PISTON DISPLACEMENT

Piston displacement is the volume that the piston displaces when it moves from TDC to BDC. The bore and stroke are used in calculating piston displacement.

Problem: An engine has six cylinders with a bore of 3½ inches and a stroke of four inches. What is the piston displacement?

The formula used to find displacement is

$$PD(Piston\ Displacement) = \frac{\pi \times D^2 \times L}{4}$$

D = diameter of bore in inches.
L = length of stroke in inches.

Substitute the figures from the problem in the formula to find the displacement of one cylinder.

$$\frac{3.14 \times (3½)^2 \times 4}{4} = \frac{153.86}{4} = 38.47\ cu.\ in.$$

The volume of one cylinder multiplied by the number of engine cylinders will give total displacement for the engine. In this case, total displacement would be 38.47 × 6 = 230.82 (cubic inches).

In the metric system the lengths of the bore and stroke are expressed in

centimeters (cm) and the piston displacement is expressed in cubic centimeters (cc). The above problem can be
solved as follows in the metric system:

> One inch = 2.54 cm
> Bore = 3.5 × 2.54 = 8.89 cm
> Stroke = 4 × 2.54 = 10.16 cm

Then:

$$PD = \frac{\pi \times D^2 \times L}{4} =$$

$$\frac{3.14 \times 8.89^2 \times 10.16}{4} =$$

630.65 cc per cylinder.

Total PD = 630.65 × 6 = 3783.9 cc
also the PD (in liters) = $\dfrac{3783.9}{1000}$ =

3.7839 liters.

THE CLEARANCE VOLUME

The *clearance volume* is the space
between the engine head and the top of
the piston when the piston is in the TDC
position.

The Combustion Chamber

This same area is referred to as the
combustion chamber. Air is taken into
the engine on the intake stroke. When
the piston reaches BDC and then starts
toward TDC, compression of the air-fuel
mixture (air only, in the diesel) begins.
This volume of air will be compressed
into the clearance volume between the
top of the piston and the engine head
when the piston reaches TDC.

THE COMPRESSION RATIO

The *compression ratio* is the number of times this volume of air is compressed. Figure 4-2 (A) illustrates this.

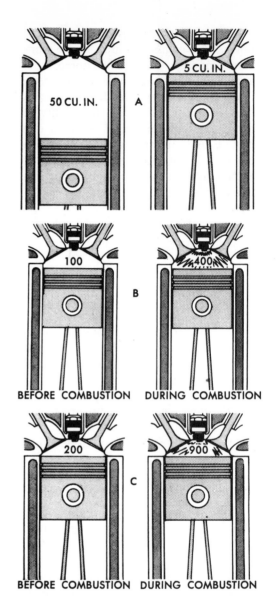

FIG. 4-2.   (A) Clearance volume (5 cubic
inches) and compression ratio. (B) Before
combustion, the cylinder pressure is only 100
pounds. As the spark plug ignites the fuel mixture, the cylinder pressure increases to 400
pounds. (C) Because the air-fuel mixture is
more highly compressed, the initial cylinder
pressure during combustion is higher, giving
the engine more power. (*Courtesy* Ethyl
Corporation)

An example will help you to understand how to determine compression ratio. The volume of a cylinder with the piston in the BDC position is 50 cubic inches, and the clearance volume is 5 cubic inches. Cylinder volume with the piston at BDC divided by clearance volume equals compression ratio. That would be 50 ÷ 5 or 10, more commonly expressed as a 10:1 compression ratio.

An air-fuel mixture reacts more violently when compressed into a small space. This is the principle of the high-compression engine. Cylinder pressure is higher in the high-compression engine than in the low-compression engine. See Figure 4-2 (B) and (C).

To understand this we must first look at a few chemical principles. We have all heard of the atom bomb or the splitting of the atom and how atomic power will be used in the future. But do we all understand what an *atom* is? The atom is generally referred to as the smallest whole particle of matter. All substances are made of a tremendous number of atoms. Any substance containing only one variety of atoms is termed an *element*. When a substance contains two or more varieties of atoms, it is termed a *compound*. All of the materials around us fall into one of these two categories. Water, for instance, is made of oxygen and hydrogen atoms and is therefore a compound. Whenever atoms combine, a *molecule* is formed. Usually a molecule is made up of many atoms.

Since all materials are made up of a great many atoms or molecules, it is important that we understand that these particles are in motion within the material. If we can visualize these molecules moving about, colliding with one another, we will find it less difficult to understand the heating and combustion of an air-fuel mixture. Heat is the rapid movement of molecules within a substance; therefore, the more rapid the movement, the greater the heat of the substance. If the heat is intense, the substance may undergo a change of state, as is the case when ice (solid state) changes to water (liquid state). Further heating or molecular motion within the substance could change the water to the vapor or gaseous state.

We know that two or more atoms can combine to form another substance. Two hydrogen atoms combining with one oxygen atom will form water (H_2O). If you have studied general science, you will remember that oxygen is essential for the burning or combustion of a material. You will also recall that heating a substance often changes it chemically. These two principles will help explain the process of combustion in an engine.

The atmosphere contains about 23 percent oxygen and 77 percent nitrogen and other gases. The fuel burned in most engines is made up of hydrogen and carbon atoms. Chemical action during combustion within an engine occurs among three elements: oxygen, hydrogen, and carbon. As the air-fuel mixture is compressed, the molecules of the mixture are forced to occupy a small space, which in turn confines the molecular activity, resulting in a tremendous collision of molecules in their effort to stay in motion. This activity creates heat which, as we know, promotes expansion. As the spark plug adds a hot spark to this compressed mixture, a violent chemical reaction takes place in which several different compounds are formed. Carbon monoxide (CO), carbon dioxide (CO_2), water (H_2O), free carbon, and nitric acid (HNO_3) are the by-products of combustion. The latter is a compound formed because of the extreme temperatures (about 4200° F) occurring in the combustion chamber. Fortunately, these by-products are in the gaseous form and can be exhausted from the engine.

The basic principles set forth in this chapter help us to understand why it is possible to obtain powerful forces from an internal-combustion engine. We must bear in mind that additional horsepower can be obtained from an engine by improving the conditions under which these chemical reactions take place.

SUMMARY

Atmospheric pressure is a very powerful force used to aid combustion and obtain power from an internal-combustion engine. Just how efficiently an engine produces power depends on many factors. Each may be likened to the function of links in a chain—if one link is weak the entire chain is weakened. It is important that we familiarize ourselves with the terms listed in this chapter. It is even more important that we understand completely the meaning of these terms and their application to our knowledge of internal-combustion engines.

Vacuum is equally important in engine operation and is a word used to indicate the absence of atmospheric pressure. The fact that areas of high pressure move toward areas of low pressure accounts for the intake of air and its important constituent, oxygen, into the engine combustion chamber. As the piston of the engine moves toward BDC, a partial vacuum is created in the cylinder. Air under atmospheric pressure then rushes into this low-pressure area, filling the cylinder with a mixture of fuel and air to be compressed and ignited. Fuel is picked up by the air as it rushes through the carburetor air passage and past the main fuel jet.

Work is the result of a force acting through a distance and can be measured by the formula: force multiplied by distance. *Horsepower,* a measure of engine power, is defined as the ability to do 33,-000 foot-pounds of work in one minute. *Power* is the time rate at which work is done. A 50-hp engine is not as powerful as a 70-hp engine, since the 70-hp engine can do more work in a shorter period of time. The 50-hp engine may be able to do the same amount of work, but the time required to do the work would be longer.

Compression ratio is the relationship of the volume of the cylinder and combustion chamber when the piston is at BDC compared to the volume when the piston is at TDC.

Carbon monoxide is a harmful by-product of engine combustion and, being an unstable compound, removes oxygen from the air to form the compound CO_2, which is a stable compound. This removal of oxygen from the air can be dangerous to human beings and animals when engines are operated inside a building with the door closed.

Several harmful gases are given off during engine operation, therefore we should exercise the greatest care when operating engines in a building or confined area.

Study the material in this chapter thoroughly, as you will find use for it many times in your study of engines.

Shop Projects

A. To demonstrate the force of atmospheric pressure, obtain a large suction cup with a handle on it. Apply the cup to a smooth, clean surface that has been moistened to obtain an airtight seal. Try to withdraw the cup. Notice the great force applied to the top of the cup by atmospheric pressure. Determine the area of the cup and compute the total force being exerted downward by the atmosphere.

B. Determine the piston displacement of an engine.

1. Use an engine that has the cylinder head removed.

2. Measure the bore in inches.

3. Measure the stroke in inches. This is the distance the piston travels in going from TDC to BDC.

4. Calculate the piston displacement (PD) in cubic inches.

$$PD = \frac{\pi \times D^2 \times L}{4}$$

5. Calculate the PD in cubic centimeters and in liters.

C. Determine the compression ratio of an engine.

1. Use a small one-cylinder engine with a vertical cylinder.

2. Compression ratio is equal to the total cylinder volume (V_1) when the piston is at BDC divided by the clearance volume (V_2) when the piston is at TDC.

$$CR = \frac{V_1}{V_2}$$
$$V_1 = \text{Piston displacement plus } V_2$$

3. Find PD from trade literature on the engine.

4. Determine V_2 by having the engine at TDC on the compression stroke and pouring a measured amount of oil into the combustion space. The volume of the oil equals V_2.

5. Now calculate CR from the above formula.

D. Determine the compression pressure of each cylinder on an assigned engine.

1. Warm up the engine before starting the test.

2. Use the compression-testing gauge.

3. Remove all the spark plugs and insert the gauge in a spark-plug hole.

4. Have the throttle in the wide-open position.

5. Turn the engine over about 10 revolutions with the starter.

6. Read the compression pressure.

7. Repeat for the other cylinders and make a chart as follows:

Cylinder No.	1	2	3	4	5	6	7	8
Comp. Press.								

If any cylinder varies over 10 percent from the average, re-test it after adding a teaspoon of oil to the cylinder.

Questions

1. Why is atmospheric pressure important to the efficient operation of an internal-combustion engine?

2. Why can a human being stand up under several tons of atmospheric pressure?

3. Why is the entry of atmospheric air into an engine of prime importance?

4. How can the piston displacement of an engine be determined?

5. What is compression ratio?

6.   How do high-compression ratios help obtain more power from engines?

7.   How can the gases given off by internal-combustion engines harm us?

8.   Define the following terms:

   (a)   work

   (b)   power

   (c)   horsepower

References

Fundamentals of Machine Operation, Tractors, Deere & Company, Moline, Illinois, 1974.

Fundamentals of Service, Engines, 3rd Edition, Deere & Company, Moline, Illinois, 1977.

Understanding and Measuring Power, Motors, Engines, Automobiles, Trucks, Tractors, SI (metric) Terms Included. 2nd Edition, American Association for Vocational Instructional Materials, Athens, Georgia, 1978.

5 Fuels and Principles of Combustion

In this chapter we will discuss fuels used in internal-combustion engines as well as the properties of these fuels. From this study we should gain an understanding of how fuels that are properly processed and compounded can bring about excellent performance, economical service, and trouble-free operation of an engine.

DISTILLATION AND PROCESSING OF CRUDE OIL

Crude oil taken from the vast oil resources in the earth is the basic material from which we obtain engine fuels. A great deal of refinement is necessary in order to break down the crude oil into its usable components. This breakdown is accomplished in a large vertical tube called a *bubble tower,* in which crude oil is heated to a point where nearly all the liquid is changed to a vapor. Since each substance has different vaporization and condensing temperatures, it is possible to draw off the various products at different levels in the tower. Figure 5-1 illustrates the principles of fractional distillation and the products obtained from this process. Additional refining is required to make these products suitable for engine use.

Gasoline is not a simple substance obtained by the process shown in Figure 5-1. The product we use is the result of blending several different hydrocarbons until just the right mixture is obtained. A well-formulated gasoline must have at least four important properties: (1) ease of starting, (2) quick engine warm-up, (3) high power potential, and (4) economical engine operation. The ease with which a fuel will start an engine depends on the volatility of the fuel.

Volatility refers to the ability of a liquid to change to a vapor. To illustrate a difference in volatility of liquids, we can use gasoline that has an initial boiling point of approximately 105° F (see Figure 5-2), as compared to lubricating oil with a boiling point of 600° F. Since gasoline will change from a liquid to a vapor at a relatively low temperature, we say that gasoline is quite volatile. Crankcase oil is not volatile, because it does not readily evaporate. Liquids will vaporize or volatilize more rapidly at high temperatures than at low temperatures.

Quick engine warm-up depends on how much of the fuel vaporizes just after the engine starts. Volatility at this point does not have to be as high as for starting the engine. It is quite evident that the need to meet these two engine demands requires careful blending of hydrocarbons of varying volatility.

High power potential and good fuel economy depend on those parts of the mixture that are low in volatility and high in heat content. Table 5-1 illustrates the difference in the heat content of the various fuels used in engine operation. It is evident that the heavy fuels produce more power per unit volume than the light fuels. Light fuels are higher in volatility and make engine starting easier, particularly at lower atmospheric temperatures.

DIESEL AND LP FUELS

Figure 5-1 shows fuel oils and gas being separated from crude oil. Diesel fuel is made primarily from the fuel-oil group and is a product of a blending process similar to that of gasoline. It has been proved that the cracking process now in common use can be applied to the fuel-oil group. This will enable us to convert more crude oil to usable gasoline. Liquefied petroleum gas (LPG), often called propane, is derived from the gas or vaporous product obtained from the top of the bubble tower. Propane is gas at ordinary temperatures and has a boiling point of −44° F at atmospheric pressures. It is apparent, then, that this fuel must be stored under pressure to keep it in the liquid state. Butane is another LP fuel that is widely used. Butane is blended into gasoline, especially for

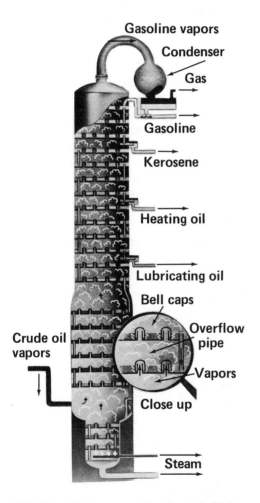

FIG. 5-1.   The process of fractional distillation and the products obtained in this process. (Fractional distillation refers to the separation of materials by a change of state from liquid to vapor.) (*Courtesy* Ethyl Corporation)

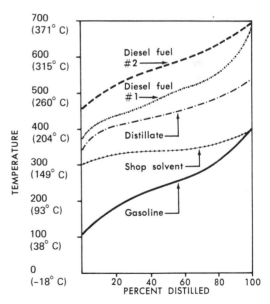

FIG. 5-2.   Typical distillation curves for engine fuels and for a shop solvent. (Drawing by Roger Cossette)

TABLE 5-1.   HEAT CONTENT, WEIGHT, AND OCTANE RATINGS OF

Fuel	Weight Per Gal. (in Lbs.)	Octane Rating	Heat Content Per Gallon (BTU)
Propane	4.25	120	93,000
Butane	4.80	100	103,000
Gasoline (Reg.)	6.15	89	125,000
Diesel Fuel	7.06	45 (Cetane No.)	139,000

* Figures shown in this table are approximate.

winter use. Below 32° F, butane is a liquid. It is not as suitable for cold weather operating conditions as propane.

HOW FUELS PRODUCE POWER

An air-fuel mixture is compressed in the combustion chamber and is ignited by the spark plug in Figure 5-3 (A). When gasoline burns in an engine, a smooth application of power is applied to the piston head in Figure 5-3 (B). The rate at which the fuel burns when ignited by the spark plug is called the *rate of flame propagation* and should be thought of as a burning action rather than an explosion. Normal flame propagation is shown in Figures 5-3 (C) and (D).

If the rate of propagation is too fast, part of the air-fuel mixture in the combustion chamber is highly compressed and will explode, rather than burn. Figures 5-3 (E), (F), and (G) illustrate the sequence that results in an explosion or knocking condition. This type of engine knock is termed *detonation* and will be heard as a steady noise, since it occurs on the power stroke of each engine cycle. This exploding action, called *knocking*, is very harmful to engine parts and results in loss of power in the engine. Improper ignition timing, a carbon buildup in the combustion chamber (resulting in an increased compression ratio), and gasoline too low in octane rating are the most common causes of engine knocking.

Preignition is another type of engine knock noticed as a result of fuel burning. This type of knocking is often termed "wild knocking" because it may occur at any time the fuel-air mixture is present in the cylinder. Since we know that temperatures of up to 4500° F are present in the combustion chamber, and that the temperature at which iron or steel will glow with a red color is approximately 1500° F, we can visualize that a valve, a spark-plug base, spark-plug electrodes, carbon flakes, or a piece of steel projecting into the combustion chamber could become hot enough to ignite the mixture before the spark plug actually fires. This situation usually occurs when engine temperature exceeds 180° F in the water jacket, or cooling system, of the engine.

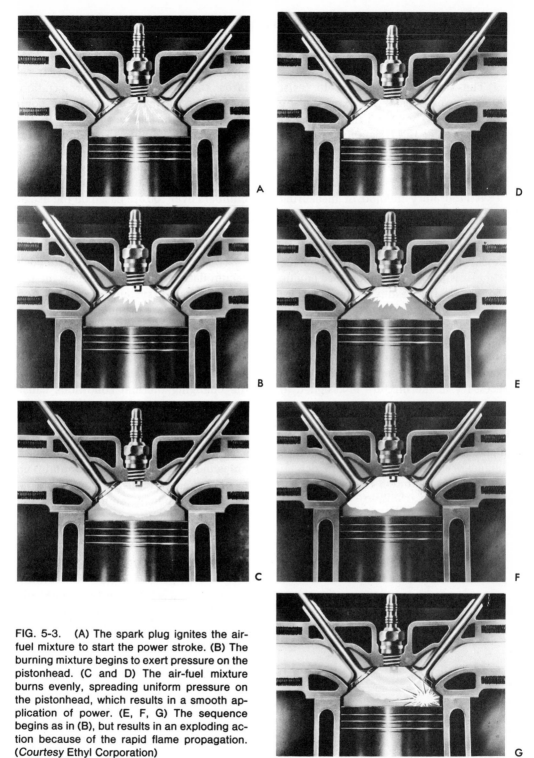

FIG. 5-3. (A) The spark plug ignites the air-fuel mixture to start the power stroke. (B) The burning mixture begins to exert pressure on the pistonhead. (C and D) The air-fuel mixture burns evenly, spreading uniform pressure on the pistonhead, which results in a smooth application of power. (E, F, G) The sequence begins as in (B), but results in an exploding action because of the rapid flame propagation. (*Courtesy* Ethyl Corporation)

PROPERTIES OF FUELS

The ability of a gasoline to resist detonation or knocking is expressed by *octane rating*. The higher the octane rating, the greater the resistance of fuel to knocking. The lower the octane rating, the less resistance the fuel has to knocking.

A laboratory method of measuring antiknock qualities of gasoline uses isooctane and normal heptane as reference fuels. When a fuel of unknown antiknock value is to be given an octane rating, it is placed in a laboratory engine and tested. The results of this test are recorded on a knockmeter. Next, a mixture of isooctane and normal heptane that will produce the same antiknock qualities as the unknown fuel is placed in the engine. The percent of isooctane necessary in the mixture is assigned to the fuel as an octane number. Suppose a mixture of isooctane (90 percent) and heptane (10 percent) placed in the test engine has the same antiknock value as the fuel of unknown quality. The octane rating of the unknown fuel would then be 90.

Knocking or detonation can be controlled by the addition of small quantities of tetraethyl lead to gasoline. This liquid lead is often called ethyl. It retards the rate of flame propagation, thereby reducing the rate of compression of the air-fuel mixture in the combustion chamber. The addition of tetraethyl lead to a gasoline raises the octane rating. Octane ratings over 100 are becoming quite common. These numbers over 100 are based on the use of isooctane and normal heptane as reference fuels and are computed by a special formula.

Fuels containing lead may deposit accumulations of lead residue in the engine's exhaust system or combustion area. This is more pronounced where an engine, such as one in a tractor, is operating at nearly constant speed. Engines that operate at varying speeds are less susceptible to lead deposits because the changing thermal conditions cause the deposits to break loose, after which they are blown from the engine.

To meet emission control standards gasolines are being offered as no-lead or unleaded fuel. These fuels contain no tetraethyl lead additive, and should be used only when recommended by the engine manufacturer.

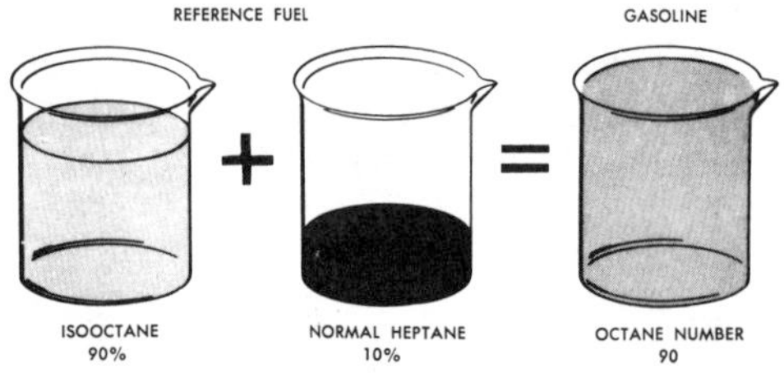

REFERENCE FUEL   GASOLINE

ISOOCTANE   NORMAL HEPTANE   OCTANE NUMBER
90%   10%   90

FIG. 5-4.   The isooctane normal heptane reference fuel combination is mixed to match the antiknock value of the unknown fuel. (*Courtesy* Ethyl Corporation)

PROPERTIES OF DIESEL FUELS

Since diesel fuel ignites with the heat of compression, it is important that this fuel ignite and burn readily at approximately 1000° F. Diesel fuel must have the proper ignition quality, or ability to burn without excessive lag. The cetane number of a diesel fuel is an indication of its ignition quality. The cetane number test is similar to the octane number test. A test engine having a variable compression ratio is used. The cetane number of the fuel being tested is determined by its performance in the engine compared to the performance of reference fuels of known cetane number. A slow-burning, low-cetane number fuel or one with a long ignition delay would knock in the engine, especially at idling speeds. Fuels with a high-cetane rating ignite readily and burn without ignition delay or lag.

Viscosity is an important property of diesel fuel. The fuel must be of a viscosity to allow it to flow readily through the pump and distribution system of the engine, as well as to atomize readily in the combustion chamber. Because the fuel acts as a lubricant for the diesel pump, viscosity must be high enough to lubricate these moving parts.

Two grades of diesel fuel are commonly available for farm tractors. Grade No. 1-D is for use in high-speed engines, such as farm tractors, where frequent and wide variations in loads and speeds are imposed upon the engine. No. 1-D is also recommended where abnormally low fuel temperatures exist.

Grade No. 2-D is less volatile than No. 1-D and is generally used in high-speed engines, such as farm tractors, that are subject to high loads and rather uniform speeds. No. 2-D is more commonly used in farm tractors.

SUMMARY

Crude oil taken from the vast reservoirs in the earth is refined to provide the fuels used in modern engines. Several other products, such as light oils, heavy oils, and fuel oils, result from the distillation of crude oil. Propane, butane, gasoline, and diesel fuel have varying heat quantities per gallon. Knowledge of this fact will help make clear why diesel engines generally use less fuel per hour than do gasoline engines.

Volatility and antiknock qualities are important in gasolines and other fuels. Sulfur and gum- and varnish-forming materials are harmful chemicals found in fuels that are not properly refined and blended. Rigid controls must be used to keep these harmful materials to a minimum during the refining process.

The fuels used in today's engines are highly refined compounds designed to give maximum engine power, long engine life, and trouble-free operation. Careful selection, correct handling, and proper storage of fuel are of prime importance to the machine operator.

Knocking of an engine is often blamed on improper octane rating of a fuel. Improperly adjusted engine timing, high atmospheric temperatures, carbon deposits in the engine combustion chamber, and a lean air-fuel mixture are other factors that cause engine knocking. Engine knocking is very harmful to engine parts and should be avoided under all conditions. Even though it is not always possible to hear knocking, particularly at high engine speeds, damage is being done to the engine when this condition exists. Always use the type of fuel recommended by the manufacturer of the particular engine that you are currently operating.

Shop Projects

A. Using Table 5-1, compute the number of British Thermal Units (BTU's) of heat that can be purchased for one cent (1¢), based on your local fuel prices for each fuel listed.

B. If an engine is available, use a low-octane fuel in it to listen to knocking or pinging. Have your instructor advance engine timing beyond specifications. Operate the engine under load to hear the knocking or pinging noise. OPERATE ENGINES FOR VERY SHORT PERIODS OF TIME.

Questions

1. Explain how crude oil is separated into its component parts.

2. Name four important properties of gasoline as a fuel.

3. Define octane rating; cetane rating.

4. Why is engine knocking harmful?

5. Name two types of knocking and explain how and why they occur.

6. What factors other than improper fuel might cause engine knocking?

7. Why is butane unsuited for cold-weather engine operating conditions?

References

Fuels & Lubricants, Selecting and Storing, American Association for Vocational Instructional Materials, Athens, Georgia, 1973.

Fundamentals of Service, Engines, 3rd Edition, Deere and Company, Moline, Illinois, 1977.

Fundamentals of Service, Fuels, Lubricants, and Coolants, Deere & Company, Moline, Illinois, 1973.

6 Fuel Systems

Gasoline and LPG fuel systems will be discussed in this chapter. Diesel fuel systems will be discussed in a later chapter on Diesel Engines.

An understanding of how each system operates will help in diagnosing fuel system problems that may occur in everyday use of an engine or tractor.

AIR-FUEL RATIOS

The gasoline and LPG fuel systems must supply the engine with a mixture of air and fuel that is combustible. It must burn in the engine when it is ignited by the spark plug.

Air-fuel ratios are expressed in terms of weight, not volume. A 12-to-1 ratio would mean 12 pounds of air to 1 pound of fuel. A mixture that is high in fuel content (9 to 1) is said to be a *rich mixture*. A high proportion of air to a small proportion of fuel (17 to 1) is said to be a *lean mixture*. It is the job of the carburetor to provide the proper air-fuel mixture for the engine. For starting purposes, the mixture must be rich (about 9 to 1) because not all of the fuel vaporizes in a cold engine. For cold-weather starting, the choke may provide a mixture as rich as 2 to 1.

As the engine warms up, the air-fuel ratio must be leaned to about 12 to 1, because a higher percentage of the fuel va-porizes in a warm engine. Medium speed air-fuel ratios should be about 15 to 1, because this ratio provides the most economical engine operation.

THE GASOLINE FUEL SYSTEM

Figure 6-1 illustrates the gasoline system. The fuel strainer is attached to the bottom of the fuel tank and is designed to remove foreign material from the fuel before it reaches the carburetor. This is a gravity system. A system incorporating a fuel pump would have the pump located between the fuel tank and the carburetor. The fuel would be delivered under pressure.

THE CARBURETOR AND ITS FUNCTIONS

The function of the carburetor is to provide the engine with the proper air-fuel ratio to meet the varying demands of engine load and speed. It must also vaporize the fuel (gasoline) and mix it thoroughly with air so that the mixture will burn properly in the engine.

If we become familiar with the parts of a carburetor, it will be easier for us to understand how the carburetor mixes and distributes the air-fuel mixture to the engine. Figure 6-2 shows a schematic

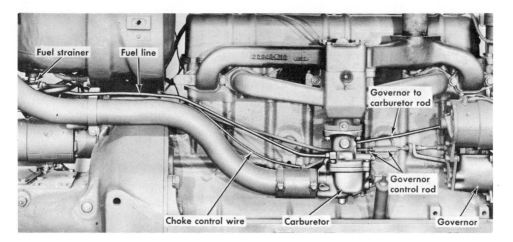

FIG. 6-1.   The gasoline fuel system showing governor connections to the carburetor. (*Courtesy* Harvester Company)

cross section of a carburetor. The numbered parts in this illustration are identified in the next several pages.

To simplify the explanation of how the carburetor functions, we will consider the four systems of the carburetor: the fuel-supply system, the idling system, the load system, the choke or starting system.

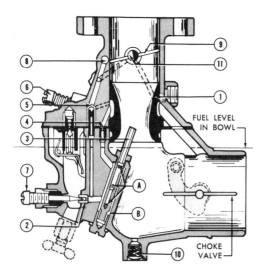

FIG. 6-2.   Parts of the carburetor. (*Courtesy* International Harvester Company)

The Fuel-Supply System

The fuel-supply system of the carburetor consists of the fuel-inlet strainer, fuel valve and seat, fuel float, fuel bowl, and bowl air vent. The function of the float and fuel valve is to maintain the correct level of fuel in the bowl under all operating conditions. The correct fuel level and proper bowl ventilation will insure that the engine gets sufficient fuel for good operation. The fuel bowl surrounds the discharge nozzle and accelerating well on three sides, thus allowing the engine to operate on angles up to 20° without flooding or affecting the air-fuel mixtures.

The Idling System

The idling system consists of the idling jet (5), (numbers refer to Figure 6-2) idle discharge slot (8), and idling well. Notice that the idle discharge slot is located above or on the engine side of the throttle plate. The fuel is drawn by engine vacuum from the idling well (adjacent to the main jet) up through the idle passage to the idling jet. Here air from the opening of the air-adjusting needle (idle-adjusting needle) (6) mixes with the fuel to make up the idle mixture. This mixture is drawn into the engine through the idling slot. Turning the idle-air-adjusting needle toward its seat reduces the amount of air admitted and increases the suction on the idle-fuel jet, which results in a richer mixture. Turn-

ing the needle in the opposite direction would admit more air and result in a leaner mixture. The idle system functions during idle or low-speed engine operation. As the throttle is opened, the throttle plate (9) opens the air passage through the venturi (1) and past the main discharge nozzle (3), causing fuel to be drawn from this source, and starts the functioning of the load system. Figure 6-4 shows the idling system in more detail.

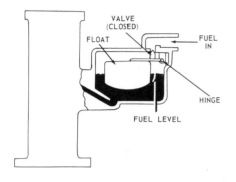

FIG. 6-3. The fuel-supply system showing the float valve, float, fuel bowl, and the fuel level. A fuel strainer is usually provided at the "fuel in" point. (*Courtesy* Deere and Company)

FIG. 6-4. The idling system. Fuel for idling enters above the throttle valve. (*Courtesy* Deere and Company)

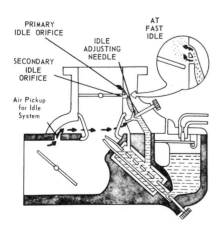

The Load System

The load system consists of the venturi (1), main jet (2), metering well, and main jet-adjusting screw (7). As the throttle plate is opened past the idling position, an increased amount of air is drawn through the venturi. The velocity of the air is increased as it passes through the restriction of the venturi, where the outlet of the main nozzle is located. The purpose of the venturi is to create a partial vacuum at the discharge nozzle. Because the float chamber and metering well are vented to the atmosphere, normal atmospheric pressure is placed on the fuel, causing it to flow through the main jet into the metering well and out through the discharge nozzle into the airstream.

The main jet (2) is a calibrated opening large enough to permit the flow of the maximum amount of fuel necessary for full-load operation. When the engine is idling, the levels of fuel in the metering well and at the discharge nozzle (3) are close to that in the fuel bowl. As the load system goes into operation, fuel is drawn from the discharge nozzle at a rate higher than that being supplied to the well by the main jet. This lowers the level of fuel in the metering well, uncovering the holes in the discharge nozzle ("A") and permitting air from the main air bleed (4) to enter the well ("B"). The addition of air is necessary to compensate for the fact that the partial vacuum produced at the main discharge nozzle increases out of proportion with the increased velocity of air through the venturi. If it were not for this metered introduction of air into the nozzle to lean the mixture, the proportion of fuel to air would steadily increase until an extremely rich mixture would prevail at full-throttle or full-load operation. The extra amount of fuel needed for rapid response on acceleration, or when the throttle is opened rapidly, is picked up from the main jet by the increased velocity of air through the venturi. This extra fuel remains above the main jet in the metering well ("B") during part-throttle operation.

An "economizer" (11) is used on some carburetors. This is not a part of the idling system. It consists of a passage from the carburetor air intake to the bowl. The passage is linked to a notch in the throttle shaft. When the throttle is between one-fourth and three-fourths open, this passage is open. It causes a reduced pressure on the bowl. This results in a leaner air-fuel mix for economy.

During acceleration up to one-fourth throttle, and for loads over three-fourths throttle, this passage is closed.

The Choke or Starting System

The choke system consists of a choke valve located in the air intake of the carburetor. When the valve is closed, air intake to the engine is restricted. This increases engine vacuum on the fuel discharge openings of the carburetor for starting purposes.

When the engine is cold, a very rich starting mixture is needed because the fuel does not vaporize easily at low temperatures and slow cranking speeds. When the engine starts, less choking is necessary to keep the engine running. As the engine warms up, the choke valve is manually or automatically opened to provide the proper air-fuel mixture.

Carburetor Adjustment

Adjustment of the carburetor should not be attempted until the engine reaches normal operating temperature. Close the throttle until the engine runs at a fast-idle speed. Turn in the idle-speed adjustment until the engine begins to increase in speed. Completely close the

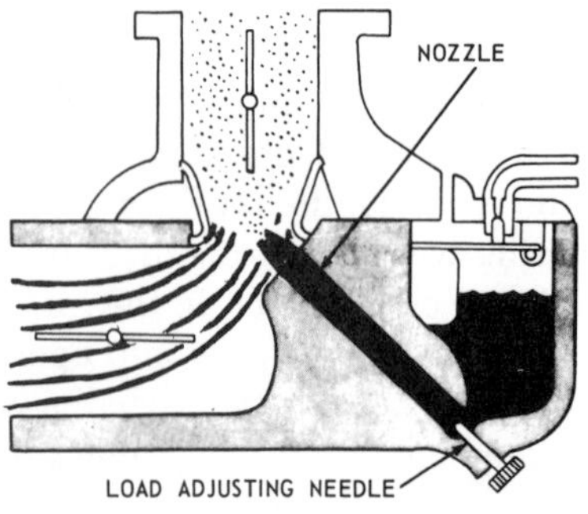

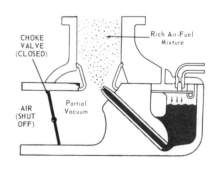

FIG. 6-5. The load system of a carburetor. (*Courtesy* Deere and Company)

FIG. 6-6. The choke system. (*Courtesy* Deere and Company)

throttle and back off the idle-speed adjustment until the desired idle speed is obtained (about 425 rpm). Open and close the throttle a few times to be sure the idle adjustment returns to the same position each time. Refer to Figure 6-7 for the approximate location of the carburetor adjustment screws.

Now turn the idle fuel adjusting screw (needle) in until the engine begins to lose speed. Note the position of the screw. Turn the idle-fuel adjusting screw out until the engine begins to decrease in speed. Turning the idle-fuel screw to a point midway between in and out positions will give optimum mixture adjustment.

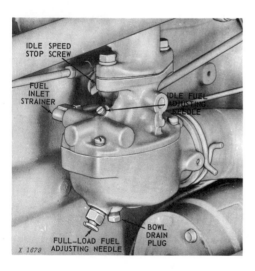

FIG. 6-7. Adjustments on a typical tractor carburetor. (*Courtesy* Deere and Company)

Carburetor Load Adjustment

Load adjustment should be made when the engine is at normal operating temperature. For maximum fuel economy, load adjustment should be made according to the work load of the engine.

If the engine is to be used on light loads, refer to the operator's manual for the initial load-adjustment setting (usually a few turns counterclockwise from the closed position). Place the engine under an average load and adjust the load-adjustment screw, using the same procedure as that used for the idle-air adjustment. For full-load operation, follow the same instructions as for part-load adjustment.

Updraft and Downdraft Carburetors

The direction of air passages through the carburetor venturi determines the type of carburetor on the engine. If the air passage is upward, the carburetor is of the *updraft* type. Air passage through the venturi in a downward direction signifies a *downdraft* carburetor. Location of the carburetor in relation to the intake manifold may also be used to determine carburetor type. Carburetors located below the intake manifold are of the updraft type; those positioned above the intake manifold are of the downdraft type. The basic principles of operation are the same for updraft and downdraft carburetors.

Manifold Heat Controls

Some engines have a manifold heat control or heat riser located in the exhaust-manifold system. The purpose of this control is to deflect hot gases in the exhaust manifold around the intake manifold to assist in vaporizing the fuel droplets entering the engine. As the engine warms up, the control gradually adjusts itself until little or no exhaust gas is deflected. This adjustment is necessary to prevent the temperature of the air-fuel mixture from rising too high. Consult the operator's manual of the engine

you are working with for proper adjustment of this control. Service the manifold heat-control mechanism regularly
with special heat-resistant lubricant to
prevent its sticking and becoming inoperative. Quicker engine warm-up, increased fuel economy, and reduced
cylinder-wall wear are reasons for servicing and maintaining the manifold-
heat control.

Fuel Level in Float Bowl

Very important to the operation and
economy of an engine is the level of the
fuel in the carburetor float bowl. The
float assembly controls the level of the
fuel admitted to the carburetor, as well
as the level of the fuel in the main jet
well. Fuel from the supply tank enters
the carburetor through the needle and
seat assembly. As the level of fuel increases in the carburetor float bowl, the
float rises, carrying the needle toward
the seat. When the float reaches the
proper level, it shuts off the fuel flow by
seating the needle valve in its seat. With
the engine in operation, fuel flows from
the bowl to the idling and load systems,
causing the float to lower and allowing
the needle to drop from its seat to admit
additional fuel. Figure 6-8 illustrates the
float bowl and regulating assembly of
the carburetor and how to make measurements when adjusting the float level.
For proper adjustment, always consult
the operator's manual.

Worn parts such as the needle and
seat, float level, float axle, and float
should be replaced if defects are noted.
Figure 6-9 shows the float assembly
parts that wear and may need replacement. The height of the fuel in the float
bowl can be determined by removing the
top of the carburetor for observation, or
by attaching a small rubber and glass

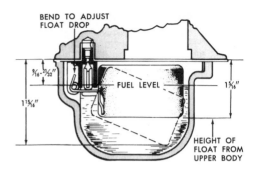

FIG. 6-8.   A typical carburetor float-bowl assembly showing adjustment measurements.
(*Courtesy* International Harvester Company)

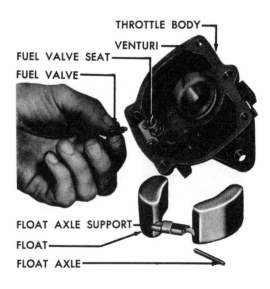

FIG. 6-9.   Parts of the float-bowl assembly
that may need replacement. (*Courtesy* International Harvester Company)

tube device to the bottom of the carburetor, and then running it upright
alongside the float bowl so that the liquid in the float bowl will seek its own
level in the glass tube. Engines having a
fuel pump should be running while this
test is being made to make sure that the
fuel in the bowl is at its normal level.

FUEL-SUPPLY SYSTEM FOR LPG ENGINES

The fuel-supply system for LPG engines differs from that of the gasoline engine. From our discussion of fuels in Chapter 5, we should recall that LPG must be stored under pressure to keep it in the liquid state. Figure 6-10 illustrates an LPG fuel-supply system, with arrows indicating the flow of fuel and air to the carburetor and manifold of the engine. As we follow the arrows, we should develop an understanding of how this fuel-supply system functions.

Note that in the supply tank of this system there are three vapor pipes extending above the level of liquid fuel in the tank. Each pipe has a labeled control valve on it. The liquid withdrawal valve has the inlet pipe located near the bottom of the tank. When full, this tank will contain approximately 80 percent liquid fuel and 20 percent vapor. It is important that sufficient volume be allowed for vapor, to handle safely the normal expansion of the liquid and vapor fuel caused by temperature changes.

Since fuel in this system is stored under pressure, it is necessary to incorporate a pressure-relief valve.

For our discussion of fuel flow and

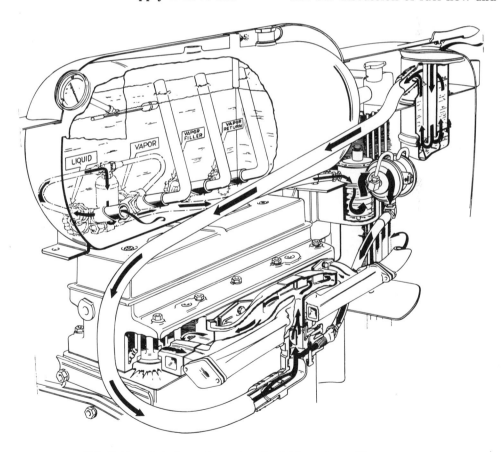

FIG. 6-10. A fuel-supply system for LPG engines. (*Courtesy* Massey-Ferguson, Limited)

engine operation of an LPG engine, we will use an illustration (Figure 6-11) showing a schematic diagram of a liquid withdrawal method of handling LPG fuel, with the parts of the system labeled for identification purposes. Following the use of this diagram, we should be able to refer back to Figure 6-10 and trace the flow of fuel and air.

For starting an LPG engine, vapor from the fuel tank must be used until the vaporizer (K) is warmed sufficiently by engine coolant to vaporize liquid fuel. The vapor-line valve (D) is opened when starting the engine, to allow vapor to be withdrawn and utilized by the engine for starting and warm-up purposes. When the engine coolant is warm enough to operate the vaporizer (the coolant temperature indicator reaches the green, or run, zone), the liquid-line

valve (F) is opened and the vapor-line valve (D) is closed. Liquid fuel passes through a filter (H) to a high-pressure regulator (J) that reduces the pressure of the fuel and changes the liquid to a vapor. From the vaporizer, the vapor passes through a low-pressure regulator (N) to the carburetor (P). The low-pressure regulator has adjustments for regulating both vapor pressure and the amount of gas that flows to the engine. The filling valve (A) and vapor-return valve (E) shown in Figure 6-11 are used when filling the supply tank. The vapor return (E) must be connected to the storage tank to allow vapor pressure to escape from the tractor tank so that liquid fuel will enter.

The vapor-withdrawal method of handling LPG fuel is used on some engines. This system has the same operat-

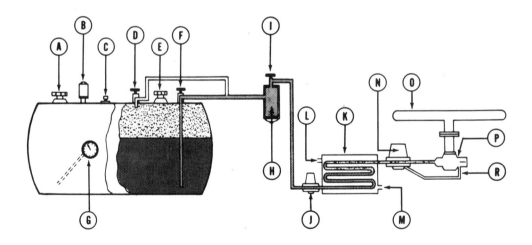

FIG. 6-11.   The parts of an LPG fuel system. (Drawing by Roger Cossette)

(A) Filling valve	(J) High-pressure regulator
(B) Pressure-relief valve	(K) Vaporizer
(C) 90-percent-full valve	(L) Hot water from engine
(D) Vapor-line valve	(M) Return line
(E) Vapor-return valve	(N) Low-pressure regulator
(F) Liquid-line valve	(O) Intake manifold
(G) Rotary gauge	(P) Carburetor
(H) Liquid filter	(R) Balance line
(I) Shutoff valve	

ing principles as the liquid-withdrawal system, except that the vaporizer is not included and all of the fuel is taken from the vapor section of the tractor supply tank.

LPG Carburetor

The carburetor of the LPG engine is bolted to the intake manifold and is sealed against the entrance of dust and dirt. The carburetor incorporates both idle and main fuel passages and has adjustments for each. Fuel is mixed with air from the air cleaner and delivered to the engine by the carburetor as it is in the gasoline engine.

Adjustment of this carburetor should not be attempted unless you have checked the operator's manual for instructions. Adjustment of the regulators should be done by a dealer or at a shop qualified to give this type of service.

SUMMARY

Storage and handling of fuel to prevent foreign material, water, and other contaminants from entering the fuel systems of our tractors are very important considerations. Purchasing a high-quality fuel from a reputable dealer can prevent fuel-system troubles. Familiarize yourself with the fuel system you are working with, in order to make sure that service and adjustments will be made properly. For proper specifications and procedures, always consult the operator's manual or other technical literature on the engine to be serviced. Special tools, gauges, and other equipment are necessary to make many of the adjustments on the LPG and diesel fuel systems.

Air-fuel ratios vary for idling, light-load, and heavy-load operation. A mixture as rich as 2 to 1 may be necessary for cold-weather starting, while approximately a 9-to-1 ratio is used at engine idle. A 15-to-1 ratio would be an economical mixture for medium-speed engine operation.

Gasoline and LPG engines use a carburetor to mix and deliver air and fuel to the engine. Apart from this fact, these two systems differ a great deal.

Shop Projects

A. Carburetor study

1. Secure a discarded carburetor from a tractor dealer or service shop. Also, get a service manual or operator's manual that shows the carburetor parts. (Figure 6-2 may be of help.)

2. Disassemble the carburetor.

3. Look for and identify the following parts: venturi, throttle valve, choke valve, main jet, main jet adjustment, idle-adjusting screw, idle jet, fuel bowl, float, float valve.

4. Are all of the above parts on the carburetor?

5. Trace the flow of fuel and flow of air through the carburetor.

6. Reassemble the carburetor.

B. Examine the fuel system of gasoline and LPG engines. Trace the fuel flow in each system, paying special attention to those parts of the system that require periodic service.

C. Carburetor adjustment

1. With the aid of the owner's manual for the selected tractor or engine learn to make the following adjustments on the carburetor.

(a) idling speed adjustment

(b) idling mixture adjustment

(c) load mixture adjustment

2. Does the carburetor that you selected have all of the above adjustments? If not, why?

Questions

1. What is the purpose of the engine carburetor?

2. Name the four systems of the carburetor.

3. List the parts of each carburetor system and give the functions of each part.

4. What is the venturi? Why is it used on a carburetor?

5. What are the components of the LPG fuel system?

6. How is liquid fuel in the propane-type tractor changed to a vapor?

7. Why is the fuel level in the float chamber of a gasoline carburetor important?

8. Why is proper carburetor adjustment so important?

References

Blue Ribbon Service Bulletins, International Harvester Company, Chicago, Illinois.

Fundamentals of Service, Engines, 3rd Edition, Deere & Company, Moline, Illinois, 1977.

Jones, Fred R., *Farm Gas Engines and Tractors,* McGraw-Hill Book Company, New York, New York, 1963.

Operator's Manual for Diesel and LPG Engines, Allis-Chalmers Manufacturing Company, Milwaukee, Wisconsin.

Tractor Maintenance, American Association for Vocational Instructional Materials, Athens, Georgia, 1975.

7 Intake and Exhaust Systems

The air cleaning system for an engine must provide large quantities of clean air. This is necessary to assure long engine life. Some of the early gasoline tractors did not have air cleaners. This resulted in severely worn engines after only a few hundred hours of operation. Modern farm engines equipped with good air cleaning systems that are properly serviced will last thousands of hours.

The air cleaner must remove dust from the air that is taken into the engine combustion chamber. A farm tractor uses about 9,000 gallons of air for each gallon of fuel that is burned in the engine. This is equal to about 164 55-gallon drums full of air for each gallon of fuel. Farm tractors generally operate in a dusty atmosphere. Dust is stirred up by the tractor as it operates in the field. Also, wind will cause dust to become air born. This dust must be removed from the air before it enters the engine. Dust is composed of small abrasive particles which will cause rapid wear if permitted to enter the combustion chamber and other engine parts.

HOW THE AIR CLEANER DOES ITS JOB

Dust can be removed from the air by several types of air cleaners. Two general principles are in use. One involves the use of centrifugal force to separate heavier dust particles from the air, and the other is an air filtering process. Many air cleaners use both principles in one unit. The job of providing clean air for the engine can be done by one of several units.

The Oil-Bath Air Cleaner

A common type of air cleaner used on older tractors is one that uses an oil-bath filtering element. This is illustrated in Figure 7-1. Many of these air cleaners are still in use. However, they generally require more frequent servicing than do the dry-type air cleaners that are common on newer equipment.

This air cleaner consists of a container that has a removable oil cup at the bottom, a screen element which fills most of the space, a central tube through which dust-laden air enters, and an outlet tube to take clean air to the carburetor. When the dust-laden air reaches the bottom of the oil cup, it has to reverse its direction to go up through the screen. Some dust leaves the air at this point. More dust is removed as the air passes through the oil-soaked screen. Oil is carried to this screen by the rush of air through the oil in the cup. This causes oil to be sprayed on the screen. There is actually a circulation of oil from the cup up to the screen and back to the cup.

FIG. 7-2. Air cleaner with the oil cup removed. This cup should be cleaned at least once a day when the tractor is in use. Refill it to the "oil level" mark with fresh oil. (*Courtesy* Deere and Company)

FIG. 7-1. Oil-bath type of air cleaner used on farm tractors. Dirty air (1), center tube (2), oil cup (3), filtering element (4), clean air to the carburetor (5). (*Courtesy* Donaldson Company, Inc.)

This circulation carries dust that has left the air at the oil-soaked screen back to the bottom of the cup. The circulation of oil keeps the screen relatively clean. This type of air cleaner is nearly 100 percent efficient in performance, if it is properly maintained. Oil of the right viscosity must be kept in the oil cup to the level indicated. The oil used should have enough viscosity to remove dust from the intake air and yet be light enough to be carried by the airstream into the screen where it will do its cleaning job. Generally, the same oil that is used in the crankcase can be used in the air cleaner. However, it is best to follow the manufacturer's recommendations.

Servicing the Oil-Bath Air Cleaner. The cup on the bottom of the air cleaner should be inspected at least once a day when the tractor is in use. If the dirt accumulation in the bottom of the cup is one-eighth to one-fourth inch deep, it should be cleaned and refilled to the proper level with new oil. See the tractor operator's manual for oil recommendations. When a high-detergent oil is used, all of the dirt may not settle out. In this case, the oil should be changed when it thickens, because this indicates that it is carrying a lot of dust and has lost its cleaning ability. Inspect the air-intake stack and clean it if necessary.

The screen element in the air cleaner needs occasional inspection. Particles of straw and other debris may accumulate on the underside of the element. These should be removed. Once a year, remove the entire air cleaner from the tractor and wash it in solvent or diesel fuel. Do not use gasoline, because it is too flammable. Make sure that the stack is clean and that all dirt is washed from the screen element. Reassemble the air cleaner and replace it on the tractor. Make sure that all connections are tight and that hoses and gaskets are in good condition.

The Oil-Soaked Element Type of Air Cleaner

The oil-soaked element type of air cleaner is used on some models of stationary engines, farm trucks, and small farm tractors. It has a fibrous filtering element which is soaked in oil. The dust-laden air has to pass through the element, and dust is removed from the air by the oil. It does not have the cleaning capacity that the oil-bath type of cleaner has and, for this reason, is not used on engines that require large amounts of air or that have to operate in dusty conditions.

This type of air cleaner is also in common use on crankcase ventilating systems. It is well suited to this purpose.

Servicing the Oil-Soaked Element Type of Air Cleaner. The oil-soaked element cleaner needs regular attention and is serviced by taking the filtering element out and washing it in a safe solvent to remove all dust particles. Shake out the excess solvent, allow the element to dry, *soak the element thoroughly in new oil,* and mount it in its proper place

FIG. 7-3.   The oil-bath air cleaner should be removed from the tractor once a year and washed in a safe solvent. This removes the dirt that has accumulated on the air-cleaner screen. (Photo by Authors)

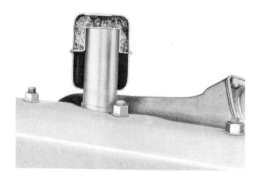

FIG. 7-4.   An oil-soaked element type of air cleaner used in an engine breather cap. (*Courtesy* Allis-Chalmers Manufacturing Company)

on the engine. The air-intake stack should be cleaned, if necessary, and all connections should be checked to make sure that they are tight. Often a good air cleaner does not get a chance to do its job because dust leaks into the air-intake system somewhere between the air cleaner and the engine. (See Figure 7-12.)

FIG. 7-5. Dry-type air cleaner used on some farm trucks and automobiles. (*Courtesy* Fram Corporation)

The Dry-Type of Air Cleaner

Several types of dry air cleaners are being used on trucks and tractors. These cleaners use a paper element, folded in accordion fashion, to filter dust from the air. Some paper elements can be cleaned and reused. Others are discarded when they become clogged with dust. Some dry-type air cleaners also use centrifugal action to remove the heavier dust particles from the air before they reach the paper element. Several dry-air ·cleaners are described below.

Figure 7-5 shows an air cleaner commonly used on automobiles and farm trucks. The paper element is chemically treated for strength and moisture resistance and is pleated to give a large filtering area. Metal screens are used to protect the paper element from possible damage by engine backfire or servicing. This type of air cleaner is easy to service and does a good job under conditions normally encountered by light trucks and automobiles.

At regular intervals, suggested by the manufacturer, the dry-type element should be removed and cleaned. Never use oil on the dry type of filter element; just shake out the dust and replace the element in the air cleaner. A new element will be necessary when the old one becomes clogged to the point where the engine loses power. Under severely dusty conditions, a new element may be necessary every 5000 miles on a farm

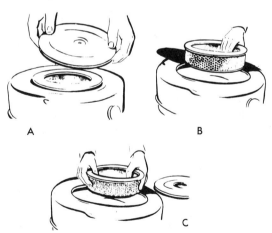

A   B

C

FIG. 7-6. A dirty dry-type filtering element can be easily replaced with a fresh element. (A) Unscrew the wing nut and lift the filter cover from the housing; (B) remove the cartridge; (C) insert the fresh cartridge into the filter housing, then replace the cover and wing nut. (*Courtesy* Fram Corporation)

truck. Under more normal conditions, this period can be extended to 15,000 or 20,000 miles. Consult the operator's manual for instructions.

A heavy-duty dry-type air cleaner designed for use on tractors is shown in Figure 7-7.

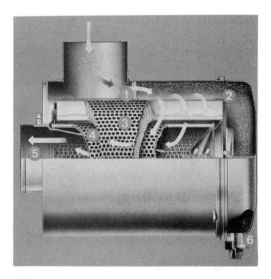

FIG. 7-7.   A heavy duty dry-type air cleaner with a precleaner (1), a primary element (3), a safety element (4), and a dust unloading valve (6). (*Courtesy* Donaldson Company, Inc.)

Dusty air taken into the system passes through angled precleaner vanes (1). This removes the heavier dust particles by centrifugal force (a swirling action). This dust is carried along the cleaner wall (2) and is conveyed to the dust cup. Dust remaining in the air is then removed by the primary filter (3). The primary filter consists of chemically treated material that is folded in accordion fashion to give a large amount of filtering surface. A perforated metal shell is added to give the unit strength and rigidity. Air passes through both the primary element and the safety element

(4). In case of accidental failure of the primary element, the safety element will protect the engine. Also, when the primary element is serviced, the safety element will filter the air. Clean air enters the carburetor at (5). The dust unloading valve (6) is provided to release dust as it accumulates from the precleaner (1).

The primary element should be cleaned and replaced, if necessary, according to the manufacturer's recommendations. Since not all dry-type elements are serviced in the same manner, it is best to follow the instructions given in the operator's manual that was furnished with the engine or with the air cleaner element. The safety element usually is not serviced, but is replaced about once each season. However, the safety element should be inspected when the primary element is serviced. In case of failure of the primary element, the safety element may become dirty, and will need to be serviced or replaced.

The rubber dust unloading valve (6) should be inspected for clogging, cracks, and deterioration. If necessary it should be replaced.

Figure 7-8 shows another dry-type air cleaner. This combines centrifugal force along with a paper element to remove dust from the air entering the engine. Dirty air enters at (1). It is given a swirling action at (2) to remove some of the heavier dust particles by centrifugal force. Most of the dust is removed by this action. The dust drops into the cup (3) at the bottom of the cleaner. The remaining fine dust particles are removed by the paper element (4). The clean air exits on its way to the engine at (5). A special dust ejection valve (6) can be added at the bottom. This valve opens (or can be opened) periodically to allow dust to fall out.

FIG. 7-9.   A vacuum gauge used to measure air cleaner restriction. When the restriction reaches the warning zone, 20 to 30 inches of water (51 to 76 cm), the air cleaner should be serviced. Refer to operator's manual for recommended restriction limit. (*Courtesy* Vortox Company)

FIG. 7-8.   A dry-type cleaner with a dust-unloading valve (6). (*Courtesy* Vortox Company)

A restriction gauge (Figure 7-9) can be added to this filter to indicate when the filter needs servicing. The gauge is graduated to show restrictions of 0 to 30 inches of water (0 - 76 cm). When the restriction of the air cleaner reaches 20 to 30 inches of water (51 to 76 cm) it should be serviced. It is best to refer to the operator's manual for the recommended restriction limit. A lower than normal reading may indicate a leak in the intake system. A sudden, sharp rise in restriction indicates a collapsed or plugged intake system.

The gauge can be mounted on the instrument panel so the operator can observe the condition of the air-cleaner element. The paper element of this type of cleaner can be washed in lukewarm water to which a nonfoaming detergent has been added. Also, dust can be removed by the careful use of compressed air. In either case the manufacturer's instructions must be carefully observed so that the paper element is not damaged. A hole in the paper element would permit dust to enter the engine.

Some heavy duty air cleaners have a continuous dust ejection feature that is activated by the engine exhaust system. Figure 7-10 shows such a system. A precleaner (1) that uses centrifugal force to remove dust is connected to a check valve and a tube (3) that leads to the engine's muffler (7). Exhaust gases rushing through the muffler cause dust to be drawn by suction from the precleaner. This dust is ejected into the exhaust

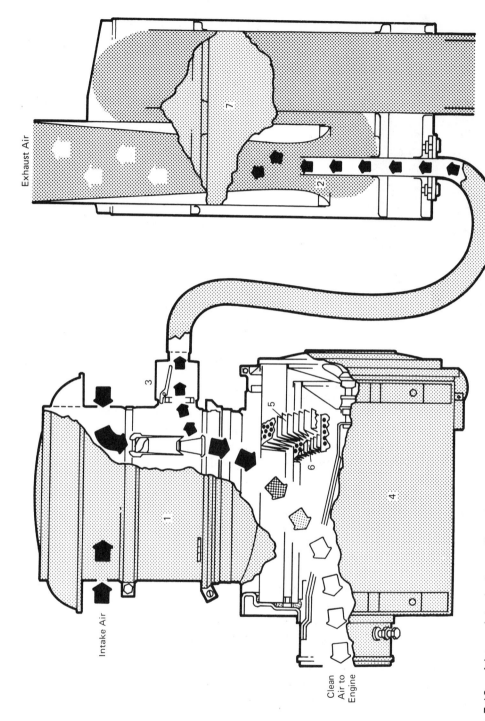

FIG. 7-10. A heavy duty air cleaner that utilizes the engine exhaust system to help remove dust. (*Courtesy Donaldson Company, Inc.*)

gases (2). The dust is then blown out with the exhaust gases. The check valve prevents backflow of exhaust gases.

Most of the dust is removed by the precleaner. The dust that is not removed by the precleaner is removed from the air by the primary element (5). The filtered air also passes through the safety element (6) before it is delivered to the engine. The safety element serves the same function as was previously described.

The precleaner (1) in this system requires little or no maintenance because dust is removed from it continuously and ejected with the engine exhaust. The primary element (5) and the safety element (6) need servicing as specified by the manufacturer.

Pre-cleaners

Pre-cleaners are also available for installation at the inlet end of the air intake system (Figure 7-11). These can be installed in an elevated position where the air is cleaner. They help remove the larger dust particles and thereby reduce the load on the main air filtering element. Also, some have screen around the air intake to prevent leaves and chaff from entering the air intake system. The pre-cleaner should be serviced by removing the outside screen and blowing or brushing off the accumulation of dust, chaff, and other dirt that has accumulated. If the pre-cleaner is neglected, then a heavier load will be placed on the main cleaning element.

MAINTAINING THE AIR-SUPPLY SYSTEM

In addition to keeping the air cleaner properly serviced, it is necessary

FIG. 7-11. Pre-cleaner. Large dust particles, leaves, and chaff are removed by this unit before the air reaches the main element. (*Courtesy* Deere and Company)

to make sure that there are no other places where dust can get into the air-intake system. A damaged hose connection between the air cleaner and the carburetor can let in enough dust to damage the engine. Keep hose connections tight and inspect the hoses periodically to make sure that there are no holes or breaks in them.

Sometimes the gasket between the carburetor and intake manifold, shown in Figure 7-12, becomes loose or may even be left out during a repair job. This will permit dusty air to enter at this point. Always use a good gasket and keep the connections tight.

On some tractor carburetors, a small piece of felt is placed in a drain hole at the bottom. This allows excess fuel to drain out. This felt must be in place to keep dusty air from entering. Worn throttle shafts can also let dusty air into the engine. These should be re-

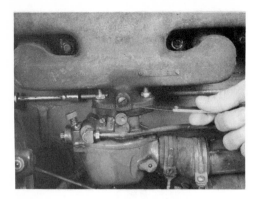

FIG. 7-12.   The gasket between the carburetor and the intake manifold was left out after an overhaul job. Dust will enter at this point and cause rapid engine wear. (Photo by Authors)

placed or repacked when they become worn.

Exhaust Systems

The burned gases must be removed from the engine. These exhaust gases flow from the combustion chamber through the exhaust valve (when the valve is open) and then through the exhaust pipe and muffler. The muffler is necessary to reduce the engine noise to an acceptable level. It also acts as a spark arrester. Two types of mufflers are in general use: the straight-through type and the reverse flow type. Both are illustrated in Figure 7-13.

The straight-through muffler has a perforated inner pipe through which the exhaust gases pass. It also has an outer shell about three times as large as the inner pipe. The space between the pipe and the outer shell is filled with material that will absorb sound and resist heat.

The reverse flow muffler has a series of short pipes and baffles that cause the exhaust gases to travel back and forth before being discharged.

The exhaust system should be kept tight and free of leaks. One of the products of combustion is carbon monoxide. This is a deadly poison, and must be kept from entering tractor cabs. Also, an engine should not be operated in a closed garage or shed without proper ventilation. Carbon monoxide is odorless and difficult to detect.

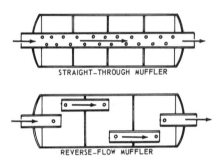

FIG. 7-13.   Two types of mufflers. Straight-through type (top) and the reverse flow type (bottom). (*Courtesy* Deere and Company)

Turbochargers and Intercoolers

The purpose of a turbocharger is to force more air into the engine than is possible with the ordinary air intake system. When more air is forced into the engine then it is also possible to add more fuel. The result is the engine develops more power. Usually it is possible to increase the power of an engine by as much as 30 percent with the use of a turbocharger. This increased power is advantageous because the horsepower of an engine can be increased without increasing the size (cubic inch displacement) of the engine. However, the addition of a turbocharger to an engine also increases engine cylinder pressure by as much as 50 percent, increases engine temperature, increases the air intake volume (a larger air cleaner may be required), increases wear on the engine and drive train, and may result in increases in oil consumption. The turbocharger is actually an air pump that is operated by the exhaust gases. Figure 7-14 shows the turbine wheel (A) which is driven by the exhaust gases and the compressor wheel (B) which is driven by the turbine wheel. The compressor wheel is located between the air cleaner and the engine intake manifold. The turbine wheel is located between the exhaust manifold and the muffler. Turbochargers rotate at speeds of 40,000 to 100,000 revolutions per minute. This high speed causes them to have a characteristic whine. Unusual noises in a turbocharger should be investigated and remedied immediately.

Turbochargers should be used only on engines that are designed for their use. Figure 7-15 shows how the turbo-

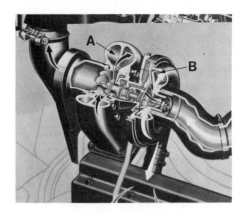

FIG. 7-14. A turbocharger. The exhaust driven turbine wheel (A) drives the compressor wheel (B). (*Courtesy* Deere and Company)

charger fits into the intake and exhaust systems. Also shown in Figure 7-15 is an intercooler. When air is compressed it

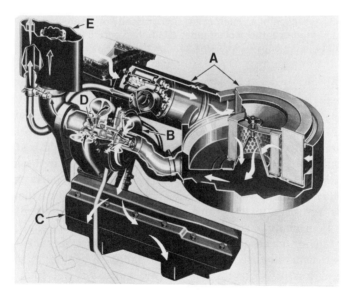

FIG. 7-15. This shows how the turbocharger fits into the air intake and exhaust systems. Air is taken into the air cleaner (A), compressed by the compressor wheel (B), cooled by the intercooler (C), and then goes to the intake manifold and engine. Exhaust gases drive the turbine wheel (D) and then escape through the muffler (E). The turbine wheel (D) drives the compressor wheel (B). (*Courtesy* Deere and Company)

becomes hot. This causes it to expand and tends to offset the purpose of the turbocharger. The intercooler cools the compressed air so that a larger volume can be forced into the engine. The intercooler is placed between the turbocharger and the intake manifold.

The intercooler is a heat exchanger.

The intake air flows over a series of tubes through which engine coolant is circulated. This reduces the temperature of the compressed air by 80 to 90°F. This results in more engine horsepower, greater economy, and quieter combustion. A cut-away section of an intercooler is shown in Figure 7-16.

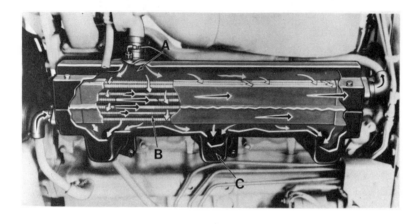

FIG. 7-16.   Cut-away section of an intercooler. Compressed hot air enters at (A). It is cooled by being passed over the coolant tubes (B). It then enters the intake manifold at (C). Coolant from the engine circulates through the coolant tubes.

SUMMARY

Clean air is essential for long engine life and good performance. The air cleaner is an important part of the engine. The oil-bath air cleaner is used on older farm tractors. Many small engines use the oil-soaked element type of air cleaner. This is also used in the crankcase breather cap of many engines.

The dry-type air cleaners with a paper element are generally used in cars, trucks, and farm tractors. The tractor-type cleaners have a centrifugal-force cleaning action in addition to the paper filtering element.

The air cleaner must be carefully serviced at regular intervals in order to permit it to do its job. When the air supply system fails to provide clean air to the engine, the engine will wear rapidly and will soon begin losing power. A little care will help avoid costly repair jobs.

The muffler is necessary to reduce engine exhaust noise and to act as a spark arrester. The exhaust system must be properly maintained so that carbon-monoxide will not enter the tractor cab.

Turbochargers are used on engines to increase engine horsepower by forcing more air and fuel into the engine than is possible without the turbocharger. Engine horsepower can be increased by as much as 30 percent with these units.

Shop Project:
Servicing an Air Cleaner

A. Oil-bath type

1. Remove the oil cup from the bottom of the air cleaner (Figure 7-2).

2. Wash the cup in fuel oil or a suitable solvent to remove all old oil and sediment from the cup. Do not use gasoline.

3. Remove any straw or trash that may be on the underside of the screen element of the air cleaner.

4. This is also a good time to inspect the air stack which goes up through the center of the air cleaner. Clean out any dirt that may have accumulated in it.

5. Refill the cup to the indicated level with new oil of the proper grade (see the operator's manual) and replace it on the air cleaner.

6. Make sure that all gaskets are in place and that all connections are tight.

7. If the tractor is equipped with a precleaner (usually a small glass jar near the top of the air cleaner stack) remove it and empty out all dirt. Then replace the jar.

B. Oil-soaked element type

1. Remove the filtering element (see the operator's manual, if necessary).

2. Wash the filtering element in fuel oil or a suitable solvent. Be sure that all dirt is removed.

3. Shake excess fuel out of the element and let it drain thoroughly.

4. Soak the filtering element in new oil of the proper grade. Let the excess oil drain off.

5. Replace the element in the body of the air cleaner and replace all gaskets, covers, and screws. Make sure that all connections are tight.

6. If the tractor is equipped with a precleaner, service it as outlined under Item 7 of the oil-bath-type cleaner project.

C. Dry-type air cleaner

1. Since servicing instructions vary for various makes of dry-air cleaners, it is best to consult the operator's manual to determine how the cleaner should be serviced.

2. Proceed according to instructions in the operator's manual.

Questions

1. Why is it important to have an air cleaner on a farm engine?

2. How does the oil-bath air cleaner do its job?

3. How often should the cup on the oil-bath air cleaner be serviced?

4. Why is fuel oil, a safe solvent, or diesel fuel recommended for cleaning the screen element of an air cleaner? Why is gasoline not recommended?

5. What are the advantages of the dry-type of air cleaner? How does this cleaner do its job?

6. How should the dry type of air cleaner be serviced?

7. Why is the oil-soaked element type of air cleaner not used on tractors that utilize large amounts of air and operate in dusty conditions?

8. Does use of a good air cleaner always mean that no dust will enter the engine? Explain your answer.

9. Why is it important to refer to the tractor operator's manual for information on servicing air cleaners?

10. How does the use of an air cleaner affect the service life of an engine?

11. What would be the effect of a restriction such as a dent in the air-intake stack?

12. What is the purpose of a turbocharger?

13. Why are intercoolers used on some turbocharged engines?

References

Fundamentals of Service, Engines, 3rd Edition, Deere & Company, Moline, Illinois, 1977.

Fundamentals of Machine Operation, Preventive Maintenance, Deere & Company, Moline, Illinois, 1973.

Fundamentals of Machine Operation, Tractors, Deere & Company, Moline, Illinois, 1974.

Jones, Fred R., *Farm Gas Engines and Tractors,* McGraw-Hill Book Company, New York, New York, 1963.

8 Valves

The material in this chapter covers the function of valves in the internal-combustion engine and the maintenance and servicing of valve assemblies.

In Chapter 3 we discussed the location of the valve assembly in the engine and illustrated the types of valve assemblies used in modern engines. Since we should now be familiar with the intake- and exhaust-valve assembly, the first item we will consider is valve arrangements commonly found in internal-combustion engines.

VALVE ARRANGEMENT

Valve arrangement refers to the location and grouping of valves to make the most efficient use of intake and exhaust manifold ports. Most engines are so arranged that two valves utilize one manifold port, except the valves on either end of the engine. These are usually exhaust valves. One reason intake manifolds are made as short as possible is to help provide a uniform distribution of the air-fuel mixture to all cylinders. Intake and exhaust valves can be identified by noting which manifold port leads to an individual valve.

Valve arrangements most commonly used are the L-head and the I-head. Engines using these valve arrangements are commonly referred to as

L-head and I-head engines, respectively. The L-head arrangement has the valves located in the engine block beside the cylinder bore. The I-head engine has the valves located in the cylinder head and is properly called an overhead-valve engine or a valve-in-head engine. Figure 8-1 illustrates valve position in L- and I-head engines.

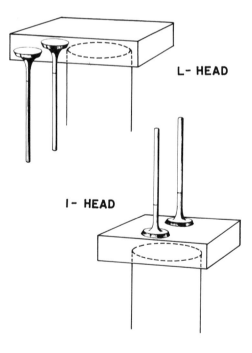

FIG. 8-1. Valve position in relation to the cylinders in L-head and I-head (valve-in-head) engines. (*Courtesy* Standard Oil Company of Indiana)

L-head engines have a simple valve mechanism. The camshaft actuates the valve through a valve lifter which comes in direct contact with the valve stem. Since the valves are located beside the cylinder bore, the combustion chamber must be large enough to accommodate the piston area as well as an area in which the valves can open and close. In the I-head engine all of the combustion space is directly over the piston. The camshaft actuates the valves through a valve lifter which in turn operates a push rod. The push rod operates a rocker arm that pivots on the rocker-arm shaft. The rocker arm comes in contact with the valve stem to produce valve action. It is evident that the I-head engine, with its complex valve mechanism, will need more frequent adjustment and maintenance than the simpler L-head arrangement. Figure 8-2 shows the rocker-arm and valve-lifter mechanisms of the two engines.

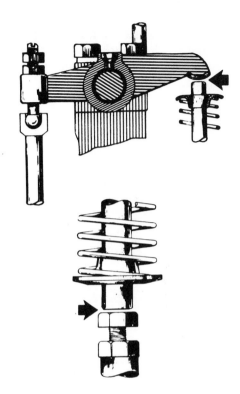

FIG. 8-2.　The rocker-arm arrangement of the I-head engine and the valve-lifter-to-valve-stem arrangement of the L-head engine. (*Courtesy Standard Oil Company of Indiana*)

VALVE TIMING

Valve timing refers to the action of the intake and exhaust valves in relation to the movement of the piston in the engine. For maximum power and performance from the engine, the valves must open and close at precise intervals in relation to piston action. If you refer again to the basic construction of engines, you will note that the camshaft is driven by the crankshaft through a timing gear or chain. Since the pistons are connected to the crankshaft through a connecting rod and the valves are actuated by the camshaft, you can see why these two mechanisms must be in time. In the four-stroke cycle engine the valve assembly must operate on two strokes of the engine cycle and be closed on the remaining two strokes of the cycle. Since the valves actually operate half of the time, it is logical that the valve train will run at one-half of crankshaft speed. This speed reduction is accomplished by placing twice as many gear teeth in the camshaft gear as in the crankshaft gear. To facilitate getting the valve train in time with crankshaft or piston action, small marks are placed on the timing gear and on the crankshaft-drive gear. These may be small punch marks, lines, or other markings. When these marks are properly aligned, the valve train and the piston assemblies are in the correct relationship and the intake and exhaust valves should open and close at the proper time. Figure 8-3 shows typical marks found on the camshaft and crankshaft gears and the alignment procedure utilized.

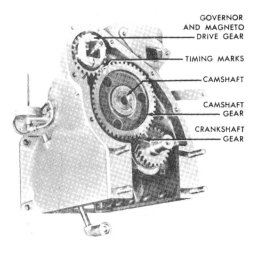

FIG. 8-3.   Proper alignment of timing marks on the camshaft and crankshaft gears. (*Courtesy* International Harvester Company)

Up to this point, we have talked of a stroke as 180° of crankshaft travel. However, the actual opening and closing of the valves vary from this pattern. Review the shape of a cam lobe on the camshaft and you will note that the valve opening and closing is a gradual process. The distance the crankshaft rotates is measured in degrees from the point where the cam lobe begins to open the valve, to the point where maximum valve lift is obtained at the highest part of the cam lobe, and then to the actual closing of the valve. The usual procedure is to begin opening the intake valve at 10° past TDC on the intake stroke and to keep it open until the piston has started the compression stroke and the crank is about 30° past BDC. Both valves must be completely closed during the compression stroke. The valves must remain closed to completely trap the force of the burning gases and to exert the greatest possible force downward on the piston head while on the power stroke. Practically all of the force of the burning gas is expended when the piston reaches a position about 135° past TDC. The exhaust valve opens at this point to allow the burned gases to be forced out of the engine on the exhaust stroke. The exhaust stroke is the longest of the four strokes, lasting about 225° of crankshaft rotation. Table 8-1 gives the usual valve timing in an engine. Note the variation in the number of degrees each valve remains open and closed in one engine cycle. Valve opening and closing points will vary from one engine to another. Figure 8-4 shows a value-timing diagram for another engine. This varies somewhat from the figures given in Table 8-1.

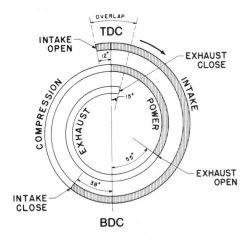

FIG. 8-4.   Valve timing for a typical tractor engine. Note that the intake opening point and exhaust closing point overlap a few degrees. (*Courtesy* American Oil Company)

TABLE 8-1   VALVE TIMING OF AN ENGINE

Stroke	Intake Valve	Exhaust Valve
Intake	Opens 10° past TDC	Closed
Compression	Closes 30° past BDC	Closed
Power	Closed	Opens 45° before BDC
Exhaust	Closed	Closes at TDC

VALVE ROTATORS

Valve rotators are used on many engines to lengthen valve life and improve valve performance. Since they are forced to work in the combustion chamber where high temperatures and chemical deposits are present, it is not difficult to understand why valves need periodic service. Rotators are mechanisms used on the lower valve stem to rotate the valve slowly during engine operation. This turning action tends to prevent valve sticking, face deposits, and valve burning. Rotation of the valve in the valve guide helps to prevent accumulation of hard cokelike deposits between the valve stem and the valve guide. It also wipes the face of the valve against the valve seat to prevent deposits from holding the valve in a partially open position, where exhaust gases will burn the valve face. Rotators can increase the length of value life appreciably, especially where combustion-chamber deposits and the length of valve life are a problem to the owner.

Valve rotators are of two basic types. The *release-type* valve rotator fits on the end of the valve stem like the ordinary valve spring retainer. It is spring-loaded. When the valve is lifted off its seat, the rotator releases the spring load, allowing the valve to turn freely in its guide. *Positive-type* valve rotators fit in the same position, but turn the valve a given amount each time the valve is lifted off its seat. Figure 8-5 (A and B) shows the two types of valve rotators and their principles of operation.

VALVE SERVICING

Valves operate under high temperature and pressure conditions. They are subjected to chemical action as a result of combustion. It follows, then, that periodic service and adjustment are necessary to give long valve life and efficient operation. We will point out the important points for you to learn about valve adjustment and lubrication, as well as the grinding and general reconditioning of the valve mechanism.

Valve Grinding

Valve grinding is the process of resurfacing the face of the valve to remove pits in the metal, to restore perfect roundness to the valve face, and to finish the valve face to its correct angle. This job can be done only on a precise machine that is properly adjusted and operated by an experienced person. The overhead-valve engine has made valve service more practical for the tractor owner. The valve-in-head assembly can be removed by the owner and taken to a

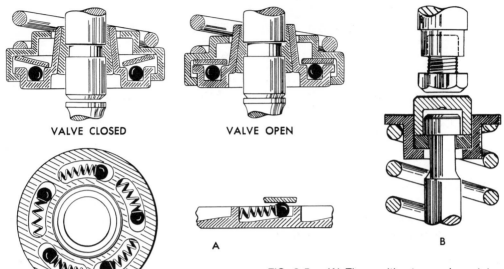

VALVE CLOSED   VALVE OPEN

A   B

FIG. 8-5.   (A) The positive-type valve rotator uses the spring-loaded ball device to turn the valve. (B) The release-type rotator merely frees the valve of its spring load, allowing the valve to turn freely. (*Courtesy* Ethyl Corporation)

service shop that is equipped and staffed to do valve work. The disassembly and reassembly work can be done in the farm shop to reduce the cost of engine service. *Valve seat grinding* is the process of grinding the valve seat in the cylinder head or block to match precisely with the finished valve. The following information will cover the complete valve-grinding job as it should be done on an engine.

Engine Disassembly. Remove the engine hood and other parts of the tractor as necessary to facilitate the removal of the cylinder head. Place the parts in order as they are removed, so that reassembly will be in proper order. The operator's manual often lists the parts to remove for valve-grinding work. Loosen and remove the cylinder-head bolts, placing all parts in a box or pan to prevent loss. Remove the rocker-arm assembly and the push rods from the valve-in-head engine, keeping the push rods in their proper order. All parts of the valve mechanism should be kept in

order so that each part is replaced in its original position. Remove the cylinder head on valve-in-head engines.

Disassembly of the Valve Mechanism. Compress the valve spring with a special tool for this purpose and remove the small half-moon keepers or other retainers from the lower valve stem. Place the valve spring, keepers, and retainer washer in order, using a device such as is shown in Figure 8-6. Before removing the valve from the cylinder head, check to see that the area near the keeper groove is free of metal burrs that would scratch the valve guide when the valve is removed. If a burr is found, remove it with a file or other tool. Remove the valves, placing them in order in the valve parts box.

Use a wire buffing wheel or wire brush to remove all foreign material from the valve. A carbon scraper or wire brush and electric drill combination

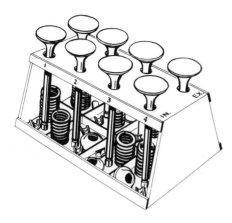

FIG. 8-6. A suitable valve-parts' box used to keep the parts in order. (*Courtesy* International Harvester Company)

FIG. 8-7. Buffing the deposits from the cylinder-head and valve-seat area. (*Courtesy* International Harvester Company)

FIG. 8-8. One method of cleaning valve guides. (*Courtesy* International Harvester Company)

should be used to clean the carbon and other deposits from the valve seat and cylinder head. Figure 8-7 illustrates this clean-up operation. Figure 8-8 shows a method used to clean the valve guides, a step to follow in the general cleanup.

Before you begin to grind the valves or valve seats, it is important that you check for valve-stem wear and valve-guide wear. This step is particularly important in valve-in-head engines, as excessive valve-stem-to-valve-guide clearance may result in high oil consumption, especially on the intake valves. Check the specifications for the engine being worked on to determine the proper valve-stem-to-valve-guide clearance. If the valve-stem or valve-guide wear is excessive, the valve or valve guide must be replaced with a new part. Measuring the valve seat out-of-roundness will tell you how much grinding is necessary to recondition this part. A thorough examination of valves, valve seats, and cylinder head for defects is necessary before reconditioning work

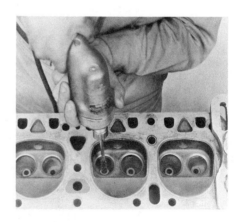

begins. Figures 8-9 and 8-10 show the measurement being made to determine the extent of mechanical wear and a basis upon which to gauge the amount of work necessary to recondition the valve mechanism.

If a valve seat is damaged beyond repair, a new valve-seat insert may be installed. Some engines will have the insert as a regular installation, while those that have integral valve seats may be repaired with an insert installation. Figure 8-11 shows the removal of a valve-seat insert.

FIG. 8-9. Measuring valve-stem wear with a micrometer. A comparison with the measurement above the wear area will give the total wear. (*Courtesy* International Harvester Company)

FIG. 8-10. Checking valve-seat out-of-roundness with a dial gauge. (*Courtesy* International Harvester Company)

FIG. 8-11. Breaking and removing a damaged valve-seat insert. (*Courtesy* International Harvester Company)

Grinding the Valve. Place the valve in the chuck of the valve-grinding machine (make sure that the chuck is set at the proper angle for this valve), being careful to align the stem in such a way that the chuck jaws grasp the valve stem beyond the worn area of the stem. This will prevent valve-head wobble and result in a more accurate job. Start the chuck and observe the rotation of the head of the valve. Wobbling of the head will indicate a bent valve stem or improper alignment of the valve stem in the chuck. Check the trueness of the grinding stone, dressing the stone if necessary. Place the valve head to be ground in line with the center of the grinding stone and slowly move the stone until slight contact is made with the valve face. Slowly move the carriage back and forth across the face of the stone until the grinding ceases. Move the stone away from the valve and observe the ground surface. If less than half the valve face shows evidence of being ground, the valve stem is probably untrue or the valve head is too far out of round to be reconditioned by grinding.

Most valves can be reconditioned by using light cuts of the grinding stone until a continuous grinding sound is heard, indicating the valve face is true. If examination shows that all pits are removed, work on that valve is completed. Do not remove any more metal than is necessary to true the valve-face surface.

If you have a mental picture of just how the valve should appear after it is ground, you can recognize those valves that should be used and those that must be replaced. Figure 8-12 shows valves that are ground correctly and those that are refaced to a point where the margin has disappeared and they are no longer serviceable.

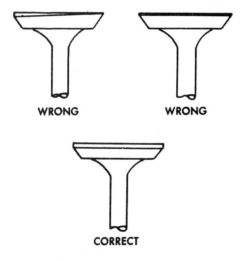

FIG. 8-12.   Valves that have been ground incorrectly (*Top left* and *right*) and one that has been ground correctly (*Bottom*). (*Courtesy* International Harvester Company)

Grinding the Valve Seat. When the valve faces are reground, you must true the surfaces of the valve seats. This is done with equipment such as is shown in Figure 8-13.

The manufacturer's guide for use of this equipment will give the necessary details on stone care, the position and use of the offset drill, and other pertinent information. Specifications regarding the use of the correct pilot stem and stone to use should also be with the valve-grinding equipment.

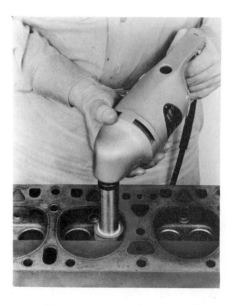

FIG. 8-13.   Grinding a valve seat. (*Courtesy* International Harvester Company)

Correct contact between valve and seat is shown in Figure 8-14. Two common mistakes made in valve-grinding work, valve seats that are either too narrow or too wide, are indicated. Special tools are available to reduce the width of the valve seat to give the correct face-to-valve-seat contact. At one time, it was common practice to lap the two valve surfaces after or instead of grinding (lapping is use of an abrasive to obtain a perfect valve-to-seat fit). This practice is no longer necessary.

FIG. 8-14. Correct and incorrect valve-to-seat relationships. (*Courtesy* International Harvester Company)

Figure 8-15 shows the reaming of valve guides to accommodate oversize valve stems. This service procedure would be used where integral valve guides are being serviced. Engines with removable guides would use new guides and standard valves.

A check of valve-spring strength should be made before reassembly of the engine. Equipment such as that shown in Figure 8-16 is not readily available in most service or farm shops, but the sight test can be used in its place. Line the valve springs up on a flat board in their proper order and observe the relative height of each spring. Springs of less than average height should be replaced or taken to a commercial shop and checked for proper strength with equipment such as that shown in Figure 8-16.

Cleanliness and accuracy are most important in valve-grinding work. Adherence to specifications for the engine being serviced is necessary.

VALVE ADJUSTMENT

After the valves and valve seats have been properly serviced and the valve mechanism has been reassembled, it is necessary to adjust the valve clearance between the valve stem and valve

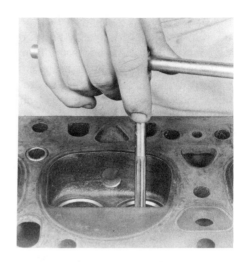

FIG. 8-15.   Reaming a valve guide. (*Courtesy* International Harvester Company)

FIG. 8-16. Valve-spring strength tester. (*Courtesy* International Harvester Company)

lifter. This clearance is essential to allow for the expansion due to heat that will tend to lengthen or expand the valve stem. When the engine reaches normal operating temperature, there must be enough clearance between lifter and stem to allow the valve spring to close the valve face flush against the valve seat. If insufficient clearance is maintained, the escaping exhaust gases will burn the face of the exhaust valves. Improper clearance on the intake valve will allow the engine to backfire through the carburetor and greatly reduce engine efficiency. The following steps may be used in valve adjustment:

1. Check the "engine hot" or "engine cold" specifications for correct valve clearance.
2. If valve clearance is to be adjusted on an engine that has just had valve service, use the cold setting.
3. If adjustment is being done on an engine in service, allow the engine to warm up to proper operating temperature.
4. Adjustment may be made on overhead-valve engines while the engine is running. Remove the valve cover.

FIG. 8-17. Adjusting overhead valves. (*Courtesy* Allis-Chalmers Manufacturing Company)

Using a gauge of the proper thickness, check to see if the gauge will pass between the rocker arm and the valve stem. Loosen the locking nut on the rocker arm and turn the screw (or nut) so as to increase or decrease the clearance to meet specifications. On L-head engines make this adjustment between the valve lifter and valve stem. Tighten the locking nut to maintain clearance. Figure 8-17 shows the tools and procedure used in valve adjustment.

SUMMARY

The most common valve arrangements in use today are the I-head (valve-in-head) and L-head types. Valve timing is the relationship between the piston action and the valve action. Precise timing of the valve system is of particular importance in obtaining maximum power and efficiency from an engine. It is also important that you become completely informed about the valve system, its condition, and its operation.

Valve condition and operation are most important to engine performance. Engine power and economy depend to a great extent on the condition of the valves. Proper valve adjustment is necessary to prolong valve life. Valve adjustment should be checked at least once during a working season, or when engine operation indicates that the valves are not functioning properly. The tractor operator's manual may suggest the intervals when this should be done. Valve rotators, by causing rotation of the valve face on the valve seat and movement of the valve stem in the guide, have eliminated many problems caused by the formation of deposits. These deposits were the cause of sticking valves and the burning of the valve face.

Careful workmanship and cleanliness, as well as adherence to the manufacturer's specifications, are important points to consider when doing valve-service work. The equipment shown in this chapter for valve- and seat-grinding work, when properly used, will add a great deal to the valve service you can expect from your particular engine.

Shop Projects

A.   Valve adjustment

   1.   Check valve clearance specifications for the engine on which you will adjust the valves.

   2.   Note whether the specification calls for the engine to be "hot" or "cold" while making the valve adjustment.

   3.   Proceed with engine "hot" or "cold" as specified.

   4.   On an engine with overhead valves the adjustment can be made while the engine is running by removing the valve cover and using a gauge of the specified thickness. While the engine is running check to see if the gauge will pass between the rocker arm and the valve stem, loosen the locking nut on the rocker arm, and turn the screw (or nut) to increase or decrease the clearance to specification. (See Figure 8-17.) (On L-head engines make this adjustment between the valve lifter and the valve stem.)

   5.   Tighten the locking nut to maintain the proper clearance.

B.   Valve grinding (This project should be done only by the more advanced student who is familiar with valve grinding equipment and techniques.)

   1.   Examine the valve-grinding equipment in a shop, observing how this equipment is operated by a qualified individual. Study the manufacturer's service and instruction manual. Become familiar with the safety factors involved in the operation of this equipment.

   2.   Secure several old valves from a garage or service shop. Practice grinding these valves.

C.   Grinding valve seats

     1.   Secure a block assembly, or cylinder head with valves, from a used auto parts dealer or ''junk'' dealer.

     2.   Familiarize yourself with the valve reseating equipment and the instructions for the equipment.

     3.   Practice grinding valve seats until you can do a satisfactory job.

     4.   Clean the valve guides in the engine head or block and check the valve-stem-to-valve-guide clearance. Compare this reading with the specifications for this engine.

Questions

1.   What are the two most common valve arrangements used in farm engines?

2.   Why is valve timing important?

3.   How is the valve train timed with the action of the pistons?

4.   At what speed compared to crankshaft speed does the valve train run? Why?

5.   How is speed reduction obtained in the valve train?

6.   What are the two common angles used between valve stem and valve face?

7.   What is the purpose of valve grinding? Seat grinding?

8.   What is valve clearance? Why must valve clearance be adjusted?

9.   Describe the process of valve adjustment on an I-head engine.

10.   Why is there a difference in appearance between intake valves and exhaust valves?

References

Fundamentals of Service, Engines, 3rd Edition, Deere & Company, Moline, Illinois, 1977.

Fundamentals of Machine Operation, Tractors, Deere & Company, Moline, Illinois, 1974.

Jones, Fred R., *Farm Gas Engines and Tractors,* McGraw-Hill Book Company, New York, New York, 1963.

9   Controlling Engine Speed–The Governor

Engines must have a speed control mechanism to provide uniform operating speeds to various machines and to prevent self destruction of the engine. Machines such as combines require a uniform operating speed. This can best be accomplished by a governor on the engine of the tractor that is pulling a pull-type combine or the governor on the engine that is driving the self-propelled combine. A governor maintains a uniform engine speed by regulating the amount of fuel that is permitted to enter the engine combustion chamber. A governor can also prevent self destruction of the engine by limiting its speed.

An engine that has no speed-control mechanism and is running with no load will run faster and faster until the engine fails or "flies apart," or until friction prevents further increase in speed. This can create very dangerous and undesirable conditions. A speed-control mechanism prevents such situations and also provides a uniform speed at the power-take-off shaft and the drawbar. Machines that are engine-driven are usually designed to operate at a certain speed. In order to maintain this speed, a speed-control mechanism is necessary on the engine.

Tractors operating up and down hills also need a speed regulating mechanism; otherwise they will travel faster going downhill and slower when going uphill. The governor maintains a uniform engine speed for all load conditions within the capacity of the engine.

The most common way to control the speed of an engine is to regulate the amount of fuel that is permitted to enter the combustion chamber. More fuel tends to increase speed at constant load or to maintain uniform speed at heavier loads. Less fuel tends to reduce speed at constant load or to maintain the same speed at lighter loads. The throttle valve in the carburetor of spark ignition engines is used to regulate the amount of fuel that is permitted to enter the engine combustion chamber. On farm tractors, this throttle valve is controlled by a governor and the hand throttle lever at the driver's seat.

On diesel engines speed is controlled by regulating the amount of fuel that is injected for each power impulse. A more complete discussion on diesel governing systems will be found in the chapter on diesel engines.

HOW THE GOVERNOR CONTROLS SPEED

Most farm engines have a governor to regulate engine speed. The most common governor is the flyweight type (Figures 9-1 and 9-2). This governor is driven directly from a rotating part of the en-

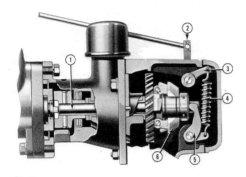

FIG. 9-1. A tractor governor showing the weights and necessary linkage. As speed increases, the weights move apart. This action moves the throttle lever towards the closed position. (1) governor shaft, (2) speed-control lever, (3) governor-control lever, (4) governor spring, (5) governor lever, (6) governor weight assembly. (*Courtesy* Allis-Chalmers Manufacturing Company)

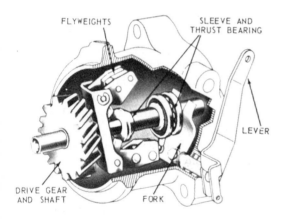

FIG. 9-2.   Cut-away view of a centrifugal governor showing the flyweights, sliding sleeve, thrust bearing, governor lever, fork, drive gear, and shaft. (*Courtesy* Deere and Company)

gine, such as the cam gear. The governor consists of a shaft, weights, springs, and a sliding collar. As the engine speed is increased, the governor-shaft speed is also increased. The increased speed and resulting centrifugal force cause the weights to move away from the shaft and stretch the governor springs. This action slides the collar on the shaft. The collar is linked to the throttle. As the engine speed is increased because of the light load, the weights separate, slide the collar, and move the throttle toward the "closed" position. This reduces the amount of fuel that is permitted to enter the combustion chamber. When the engine speed drops because of a heavier load, the springs pull the weights together because there is less centrifugal force at lower speeds. This causes the collar to slide in the opposite direction, which moves the throttle toward the "open" position, thus permitting more fuel to enter. When the spring tension and the centrifugal force of the weights are balanced, the engine will run at uniform speed.

The flyweight governor is usually lubricated from the engine's lubrication system. It requires little attention and will give good service for many years. A new governor will sometimes give trouble because of sticking parts or paint interfering with its operation. This causes the engine speed to rise above or drop below governed speed as the governor seeks its correct speed. This is called "hunting." Thorough lubrication and the removal of excess paint will usually remedy this difficulty. Governor "hunting" is sometimes caused by an incorrect (too rich or too lean) carburetor adjustment. Correctly adjusting the carburetor will remedy this.

Some air-cooled engines use the vane-type governor (Figure 9-3). This governor also regulates the throttle opening. The vane is exposed to an air blast from the engine-cooling system. As the engine picks up speed, the air blast against the governor moves the vane and the throttle toward the throttle "closed" position. As engine speed drops, a light spring pulls the vane in the opposite direction and the throttle is moved toward the "open" position. When the spring tension and the force of the air blast against the vane are balanced, the engine will run at a uniform speed. A change in governor-spring tension adjustment will change the speed at which the engine will run. The vane governor is suitable for small air-cooled engines of the type used on lawn mowers and similar machines. It governs at all engine speeds.

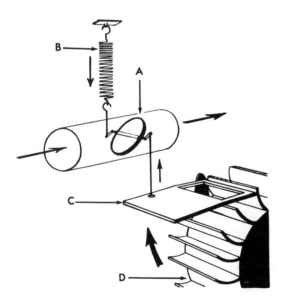

FIG. 9-3. Vane governor. Air from the flywheel (D) forces the vane (C) to move in the direction of the arrow. This action tends to close the throttle valve (A). The spring (B) counteracts the force of air on the vane and tends to open the throttle. A balance between these two forces causes the engine to run at a uniform speed. (Drawing by Roger Cossette)

SPARK IGNITION ENGINE GOVERNORS

On a farm tractor, it is essential to have governor control at all engine speeds. The governor should function when a lower engine speed is desired as well as when a high engine speed is desired. Also, it should govern at all intermediate speeds. A governor that will function at all engine speeds is known as a variable speed governor. Governing at all engine speeds can be accomplished by connecting the hand throttle lever (1) and control rod in Figure 9-4 to the governor spring (2). With this arrangement, an adjustment of the hand-throttle control lever changes the tension in the governor spring and thereby changes the governed speed. An increase in the spring tension will increase the governed speed; a decrease in the tension will decrease the governed speed.

A diagram of a centrifugal variable speed governor in operation is shown in Figure 9-5. In the left diagram the operator is opening the throttle. This causes the engine to speed up, but it also causes the governor weights to spread apart due to the centrifugal force that is created by the higher speed. However, the action of the weights spreading apart causes the governor arm to close the throttle. (See the right diagram in Figure 9-5.) Also note that the governor control spring has been stretched. When the tension in the governor control spring balances the centrifugal force of the flyweights, the governor is in equilibrium and it will govern at that speed. Moving the throttle lever merely changes the tension in the governor control spring and also changes the speed at which the governor will govern.

For a discussion of diesel governors, see the chapter on diesel engines (Chapter 12).

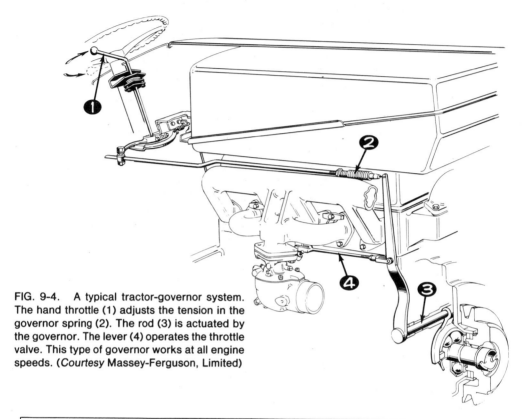

FIG. 9-4.   A typical tractor-governor system. The hand throttle (1) adjusts the tension in the governor spring (2). The rod (3) is actuated by the governor. The lever (4) operates the throttle valve. This type of governor works at all engine speeds. (*Courtesy* Massey-Ferguson, Limited)

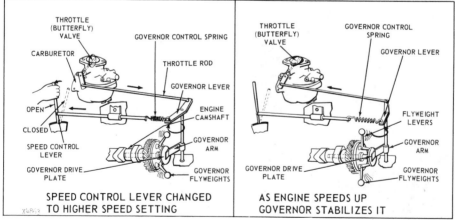

FIG. 9-5.   This shows the operation of a flyweight (centrifugal) type of governor for a spark ignition engine. (*Courtesy* Deere and Company)

MEASURING THE SPEED OF AN ENGINE OR OF A SHAFT ON A MACHINE

It is helpful to know how to measure the speed of an engine or that of a shaft on a machine. Many tractors are equipped with hour meters (Figure 9-6).

FIG. 9-7. A revolution counter and stop watch being used to determine the speed of a shaft. (A) shaft, (B) dial. (Photo by Authors)

FIG. 9-6. Engine hour meter, which also shows the engine rpm, power-takeoff rpm, and ground speed in mph. (*Courtesy* Deere and Company)

FIG. 9-8. A tachometer being used to determine the speed of a shaft. (Photo by Authors)

SUMMARY

These also have a tachometer that indicates engine speed or revolutions per minute (rpm). However, many tractors, engines, and other machines do not have such a device. It is a simple procedure to measure rpm with a revolution counter (Figure 9-7) or a direct reading tachometer (Figure 9-8). Shaft "A" on the revolution counter turns 100 times for every time that dial "B" turns around once. The counter and the tachometer can be used on any shaft that has an exposed end. A watch is used to time the revolution counter.

An engine must have a speed-control mechanism to make it possible to get a uniform and controlled speed to perform various farm operations. A flyball governor is the most widely used for this purpose. The governor spring on a tractor engine is usually connected to the hand throttle control lever so that the governor will control engine speed at all throttle settings. Engine or shaft speed (rpm) is measured with a tachometer or a revolution counter. Knowing how to measure the speed of a shaft is often useful. It is also important to run machines at their proper speeds so that they will perform satisfactorily.

Shop Projects

A.   Determining the speed of a shaft with a revolution counter

1.   Select a piece of equipment that has a shaft with an exposed end. Be sure that no keys, belts, or other items that may be dangerous will interfere with measuring the speed of the shaft.

2.   Familiarize yourself with the revolution counter. Note that it will measure shaft speeds for both clockwise and counterclockwise rotation.

3.   Use a watch with a sweep second hand or a stop watch for timing the counter.

4.   Hold the revolution counter on the end of the turning shaft and note the time (or start the stop watch) when the counter reads zero. Count the revolutions of the dial for one minute and remove the counter immediately when one minute is up. Also note the part of a revolution indicated on the dial.

Example:
If the dial on the counter turned eight full revolutions plus half a revolution in exactly one minute, the shaft speed was 850 rpm.

5.   Repeat this several times until you get the same result each time you try. A little experience is necessary in order to get the timing correct.

B.   Determining the speed of a shaft with a tachometer

1.   Place the end of the tachometer on the exposed end of the rotating shaft. (Be sure that you have selected the proper tip for the tachometer.) Note that the tachometer gives a direct reading in rpm and does not need to be timed with a watch.

C.   Checking the speed of a tractor power-takeoff (PTO) shaft

1.   Select a tractor that has a PTO shaft and an hour meter with a tachometer. (The tachometer indicates rpm.)

2.   Start the engine and increase the speed until the correct PTO shaft speed is indicated on the tachometer. Most tractors have tachometers that are marked to indicate when correct PTO shaft speed has been reached.

3.   Use the revolution counter on the PTO shaft and determine its rpm. The correct PTO shaft speed is either 540 or 1,000 rpm; some tractor PTO shafts can be made to run at either speed. The operator's manual should be consulted for the exact figures. Do your results agree with the speed given in the operator's manual?

Questions

1.   Why is it necessary to have a speed-control mechanism on an engine? Give several reasons.

2.   How is the speed of an engine usually controlled?

3.   What is the principle of a common type of governor used on a gasoline engine?

4.   Why is it desirable to govern tractor engines at all normal operating speeds? How is this accomplished?

5.   What is governor "hunting"? What causes it?

6.   What is a tachometer?

7.   What is a revolution counter?

8.   How is the revolution counter used to measure the speed of a shaft?

9.   What should the speed of a PTO shaft be in rpm?

References

Fundamentals of Service, Engines, 3rd Edition, Deere & Company, Moline, Illinois, 1977.

Fundamentals of Machine Operation, Tractors, Deere & Company, Moline, Illinois, 1974.

Jones, Fred R., *Farm Gas Engines and Tractors,* McGraw-Hill Book Company, New York, New York, 1963.

Tractor Maintenance, American Association for Vocational Instructional Materials, Athens, Georgia, 1975.

10 Igniting the Fuel Charge

In engines that burn gasoline and LP gas, the fuel charge is ignited by an electric spark. These engines are known as spark ignition engines. In diesel engines the highly-compressed hot air ignites the fuel charge. Diesels are known as compression ignition engines. We will discuss diesel engines in Chapter 12.

Spark ignition engines use an electric ignition system to provide sparks at the spark plug at the proper time to ignite the fuel charge.

Two kinds of electric ignition systems are in common use, the *battery system* and the *magneto system*. The battery system transforms the relatively low voltage of a storage battery into a high-voltage current that will jump the gap at the spark plug and thus ignite the fuel charge. In the magneto system, the current generated by the magneto is also transformed to a high-voltage current that will jump the gap at the spark plug.

The electric ignition system depends upon the principle of electromagnetic induction to do its work. In permanent magnets like those in Figure 10-1, a magnetic field exists between the north and the south poles. This magnetic field consists of invisible magnetic lines of force. If we move a wire up and down in this field (Figure 10-2), the wire "cuts" the lines of force.

This action causes a small current to flow in the wire. If we reverse the procedure and pass the electric current through a wire as shown in Figure 10-3, an electromagnetic field forms about the wire. This field can be made stronger by adding many more turns of wire and inserting a soft iron core to conduct the magnetic lines of force. (See Figure 10-4.)

A strong electromagnetic field is produced when current such as that supplied by a 12-volt battery is passed through the wire. The soft iron core becomes a strong electric magnet. If we suddenly break this wire circuit, the electromagnetic field collapses and the winding around the soft iron core actually cuts the magnetic lines of force of the collapsing field. This causes a high-voltage current to be induced in the winding and a spark to jump at the gap where the wire was broken. This is the beginning of an ignition system and is called its primary winding. The voltage produced by the system, about 250 volts, is not high enough to cause a spark to jump across the gap of a spark plug; therefore, an additional winding is added. This is called the secondary winding (Figure 10-4). It, too, cuts magnetic lines of force when we break the primary circuit and a much higher voltage is induced in it because it has many more turns of wire then the primary circuit. The voltage of the secondary circuit is 15,000 to 20,000 volts, which is sufficient to jump the spark-plug gap when

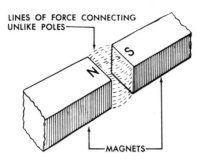

FIG. 10-1.   Magnetic lines of force between two unlike poles of magnets. (*Courtesy* Fairbanks, Morse & Co.)

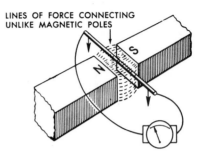

FIG. 10-2.   When a conductor making a complete circuit is moved in a magnetic field, an electric current is induced in the conductor. (*Courtesy* Fairbanks, Morse & Co.)

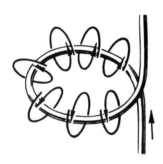

FIG. 10-3.   A magnetic field surrounds a wire when a current is passed through the wire. (*Courtesy* Fairbanks, Morse & Co.)

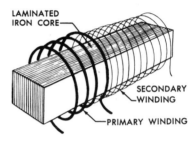

FIG. 10-4.   Both the primary and secondary windings are wrapped around the iron core to form the ignition coil. (*Courtesy* Fairbanks, Morse & Co.)

the air-fuel mixture in the engine is under compression. Both battery and magneto systems use this principle of electromagnetic induction.

BATTERY-IGNITION SYSTEM

Many farm tractors are equipped with battery-ignition systems. The battery-ignition system has a primary circuit and a secondary circuit. The primary circuit of the ignition system consists of the following parts (See Figure 10-5):

A. The battery which is grounded at the left.

B. Switch.

C. Primary winding of coil. This winding is actually over the soft iron core (center), but for clarity is shown beside the core.

L. Insulated breaker point.

M. Grounded breaker point (grounded at G). The breaker points are opened by the cam E at the proper time to provide spark at the spark plugs.

F. Condenser. The condenser absorbs electric current as the breaker points open. This reduces arcing at the points and also helps break down the magnetic field more rapidly so that a hot spark will occur at the spark plugs.

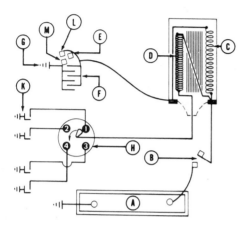

FIG. 10-5. A battery-ignition system. (Drawing by Roger Cossette)

The secondary circuit has the following parts:

D. The secondary winding which is, in practice, wound over the soft iron core (center) but for clarity is shown beside the core.

H. The distributor which distributes the high voltage spark to the proper plug at the right time. The rotor of the distributor is shown making contact with the terminal that will take the spark to number one cylinder through the wires shown.

K. The spark plugs. The center terminal on the spark plug receives the high voltage current. It then jumps the spark plug gap, ignites the fuel charge, and is grounded as shown.

This system will produce a spark each time the cam (E) opens the breaker points. As the points open, the magnetic field collapses and a high-voltage current is induced in the secondary circuit. This current takes the path of least resistance, which is across the gap at the spark plugs. This spark ignites the fuel charge. An ignition system will produce a spark in a split second. A four-cylinder engine running at a crankshaft speed of 1800 revolutions per minute requires 60 sparks per second.

An ignition system must have a mechanism to advance and retard the spark that ignites the fuel charge. When starting the engine the fuel charge should be ignited when the piston is at TDC at the end of the compression stroke. After the engine has started, the point at which the fuel charge is ignited must be reached earlier. A spark-advance mechanism that automatically times the spark to the speed of the engine is shown in Figure 10-6. This unit is located under the breaker-point plate in the distributor housing. It consists of a set of weights, springs, and a linkage to the cam. When the engine is turning at cranking speed, the springs hold the weights together and hold the cam in the "no advance" or retard position. When the engine is operating at higher speeds, the centrifugal force of the advance weights overcomes the spring tension and causes the cam that opens the breaker points to be advanced. The advance is proportional to the engine speed. The full-advance position varies with engines, but is about 25° to 30° for most engines.

Transistorized Ignition Systems

Several kinds of transistorized ignition systems are available for automo-

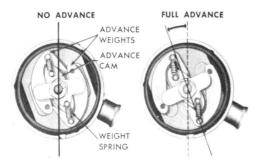

NO ADVANCE   FULL ADVANCE

ADVANCE WEIGHTS

ADVANCE CAM

WEIGHT SPRING

FIG. 10-6. Automatic spark-advance mechanism for a battery-ignition system. (Courtesy Delco-Remy Division, General Motors Corporation)

biles and tractors. The system shown in Figure 10-7 uses regular breaker points but has a special ignition coil and an amplifier. The electrical-current flow through the breaker points of this system is only a part of the current that flows through the breaker points in conventional ignition systems. The transistorized amplifier and the special coil provide a hot spark over a wide range of

operating conditions. The manufacturer lists the following advantages of this system:

1. Longer breaker-point life because of the lower breaker-point current.
2. Better starting ability with wet or fouled spark plugs.
3. A higher secondary voltage over the engine-speed range.
4. Longer spark-plug life, due to improved ability to fire wet and fouled plugs.

Servicing the Battery-Ignition System

An ignition system will function well for hundreds of hours, provided that its moving parts are properly adjusted and lubricated. All wire connections must be tight and free from corrosion. Battery connections must be clean and tight.

Breaker points become pitted and corroded. They can be cleaned with a

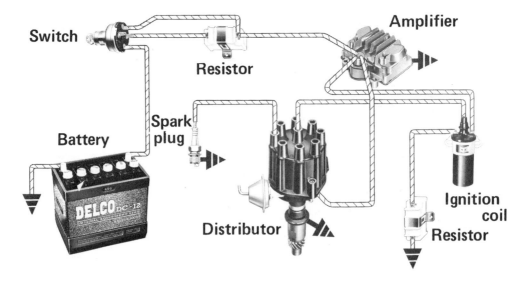

Switch

Amplifier

Resistor

Spark plug

Battery

Distributor

Ignition coil

Resistor

FIG. 10-7. A breaker-point-controlled transistorized ignition system. (*Courtesy Delco-Remy Division, General Motors Corporation*)

special point file made for this purpose. Badly pitted and worn points do not give satisfactory service and should be replaced. The condenser should be checked when points are pitted or welding has taken place. The breaker-point gap setting must be in accordance with the manufacturer's specifications. Refer to the engine operator's manual for the correct gap settings. Figure 10-8 shows

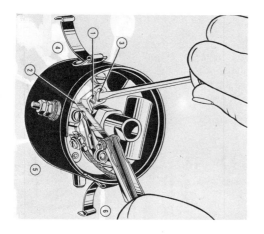

FIG. 10-8. Adjusting the breaker-point gap. To make the correct adjustment, loosen the lock screw (1), use the feeler gauge (2) to measure the gap, and make the necessary gap adjustment by turning the screw (3). See the operator's manual for the correct gap. Tighten the screw (1) after the proper adjustment has been made. The stationary point is at (4), the movable point at (5). At (6) is the hinge for point (5). (*Courtesy* Cockshutt Farm Equipment, Limited)

how to measure the gap on new points. The points must be in wide-open position when the gap is measured. An accurate adjustment is necessary because improper gap setting will affect ignition timing.

As the points become pitted and worn, the method shown in Figure 10-8 is not accurate because the feeler gauge measures the high spots of the points and will not give an accurate measurement. It is best to use a dial indicator as shown in Figure 10-9.

The number of degrees that the breaker cam rotates from the point where the breaker points close to where they open again is called the cam angle or dwell of the breaker points. (See Figure 10-10.) An increase in the cam angle will cause a decrease in the breaker point gap and vice versa. A dwell meter should be used to set the specified cam angle.

MAGNETO-IGNITION SYSTEM

Magneto ignition systems are used on many single cylinder engines used on lawn mowers, garden tractors, snow

FIG. 10-9. Checking points with a dial indicator. (*Courtesy* Deere and Company)

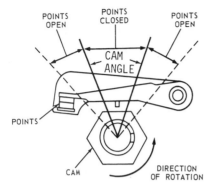

FIG. 10-10. Cam angle or dwell of points. (*Courtesy* Deere and Company)

blowers, and other small equipment. Magnetos were also used on many multicylinder engines on farm tractors. Some of these are still in use.

A magneto is an ignition system within itself, except for the spark plugs and wires leading to the plugs. Nearly all of the parts shown in the battery-ignition system can be found in the magneto. The magneto generates its electric current by electro-magnetic induction, and then increases its voltage in much the same way as in the battery system. Magnetos are equipped with permanent magnets that rotate with the magneto shaft (Figure 10-11). When the magnets rotate, the magnetic field also moves and reverses direction. This causes a current to be induced in the primary winding (Figure 10-12). The flow of current changes direction on every half-revolution of the magneto shaft. The flow is greatest just before the change in direction takes place. When the magneto is turned by hand, this point can be determined because the greatest "drag" or resistance to turning will be noted. The opening of the breaker points must be timed to coincide with the greatest

"drag" of the magneto, because this is the point where the hottest spark will be produced. As the points are opened in the primary circuit, the flow of current stops, the magnetic field collapses, and a high-voltage current is induced in the secondary circuit. This is the current that jumps the gap at the spark plug and ignites the fuel charge. The condenser in the primary circuit helps to break down the magnetic field by absorbing the pri-

FIG. 10-11. Magneto rotor with permanent magnets built into the rotor. (*Courtesy* Fairbanks, Morse & Co.)

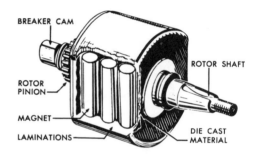

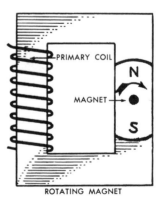

FIG. 10-12. The arrangement of the magnet and primary coil in a typical tractor magneto. As the magnet revolves, the magnetic field reverses direction and induces a voltage in the primary circuit. (*Courtesy* Fairbanks, Morse & Co.)

mary current and protects the breaker points by reducing arcing at the points.

A close examination of the magneto ignition system in Figure 10-13 will show that it is very similar to the battery system in Figure 10-5. In both systems you will find a coil (primary and secondary windings), breaker points, a condenser, a distributor, a rotor, wires, and spark plugs.

The magneto will not produce a hot spark when turning slowly, as at cranking speed. Most tractor magnetos have a device to overcome this difficulty. It is called an *impulse coupling* (See Figures 10-14 and 10-15). This coupling fits between the magneto and its drive unit. At slow speeds, the impulse coupling permits the magneto to stop turning and a spring is wound up for part of a turn or about 25 degrees of crankshaft travel. It is then suddenly released and the spring

action causes the magneto to catch up rapidly with the engine and produce a hot spark at the same time. When the engine starts running, the impulse coupling automatically disengages and the magneto turns smoothly with the engine. The impulse coupling also serves to retard the spark for starting so the engine will not kick back. When the engine starts, the spark is automatically advanced because the coupling does not function at speeds over 200 to 300 rpm.

The Flywheel Magneto. Many single cylinder engines have the magneto incorporated with the flywheel. (See Figure 10-16.) The magnets are built into the aluminum flywheel. The cam to open the breaker points is a flat spot on the crankshaft where the flywheel is attached. This magneto also contains the parts shown in Figure 10-13 except that a distributor is not required on a single

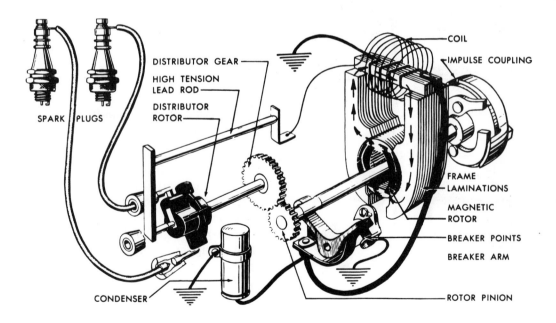

FIG. 10-13. A schematic diagram of a typical magneto-ignition system. Note that this system has many of the same parts as the battery-ignition system shown in Figure 10-5. (*Courtesy* Fairbanks, Morse & Co.)

cylinder engine. Also, the impulse coupling is not used on modern single cylinder engines.

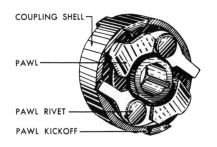

FIG. 10-14. An impulse coupling. It allows the magneto shaft to stop for a brief interval and suddenly catch up with the engine. In catching up the magneto turns rapidly and produces a hot spark for starting. (*Courtesy* Fairbanks, Morse & Co.)

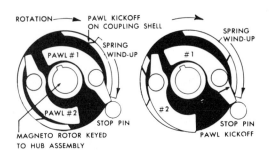

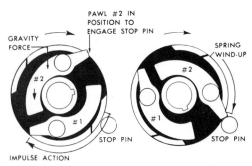

FIG. 10-15. The sequence of operations of an impulse coupling for a magneto. (*Courtesy* Fairbanks, Morse & Co.)

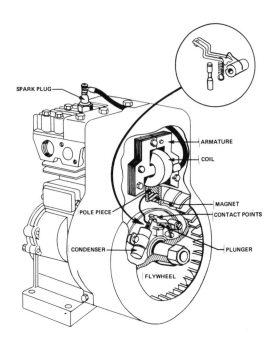

FIG. 10-16. A flywheel magneto. (*Courtesy* Briggs and Stratton Corp.)

Servicing the Magneto-Ignition System

The magneto must be handled carefully since it is an intricate mechanism with accurately fitted bearings and parts. Only minor adjustments should be made on the farm. Major repair work on a magneto should be done at a service shop equipped to handle this type of work.

The breaker points are adjusted in a manner similar to that described under the battery-ignition system. Make certain that the points are clean and have the proper gap indicated in the operator's manual (see Figure 10-17).

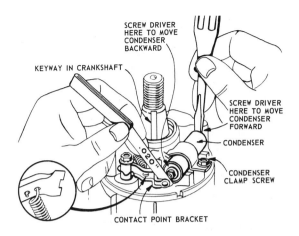

FIG. 10-17.   Adjusting the breaker points on a flywheel magneto. (*Courtesy* Briggs and Stratton Corp.)

SPARK PLUGS

The spark plug ignites the fuel charge. When the high-voltage current in the secondary circuit jumps the spark-plug gap, a spark occurs. This spark ignites the fuel charge. The spark plug must be heat resistant and well insulated. Its construction is shown in Figure 10-18.

Not all spark plugs are alike. They are made in several sizes and for use in engines that have different operating characteristics. A gasoline-burning engine may require a different spark plug than the same engine equipped to burn tractor fuel. The type of use to which the engine is subjected will also have some effect on the type of spark plug that is recommended by the manufacturer. Spark plugs are made in several heat ranges. There are hot and cold plugs and other heat ranges between these two extremes. Figure 10-19 illustrates hot and cold plugs. The hot plug has a long porcelain or insulation from which the heat cannot escape as rapidly as from a plug

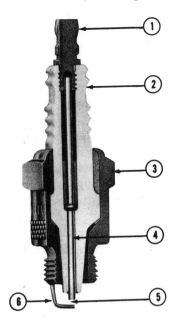

FIG. 10-18.   Cutaway spark plug showing the principal parts: terminal nut (1), insulator (2), shell (3), insulated electrode (4), spark gap (5), and grounded electrode (6). (*Courtesy* Champion Spark Plug Company)

FIG. 10-19.   Spark plugs are available in several heat ranges. (*Courtesy* Ethyl Corporation)

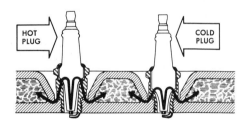

with a short porcelain. The latter is termed a cold plug.

Hot plugs are used in engines that operate under light loads. They are needed under these conditions because they do not tend to foul as readily as do the colder plugs. Cold plugs are used in engines that tend to run hot or operate under heavy loads. Cold plugs will last longer and perform better under these conditions than hot plugs. The spark plug heat ranges between hot and cold are for use in engines that operate under more normal conditions. Follow the manufacturer's recommendations to select the proper plug for an engine.

The spark-plug gap, usually from .025 to .030 inches, should be set according to manufacturer's specifications. The gap should be checked with a wire gauge as shown in Figure 10-20. A flat gauge will not give a true reading.

During normal operation spark plugs become coated with a brown or light gray deposit. This should be removed whenever necessary, which will usually be after every 250 hours of operation.

FIG. 10-20.   A plain flat feeler gauge cannot accurately measure the true width of the gap (*left*). Use a round gauge (*right*). (Drawings by Roger Cossette)

LEAD-ACID BATTERIES

The source of energy for the battery-ignition system is the lead-acid battery. A chemical reaction which liberates electricity takes place when a circuit is completed between the poles of a battery.

A battery consists of positive plates,

negative plates, separators, an electrolyte, and a case or container.

When a lead-acid battery is charged, the positive plates consist of lead peroxide (PbO_2), the negative plates are sponge lead (Pb), and electrolyte is dilute sulphuric acid (H_2SO_4) having a specific gravity of about 1.280. This means that the dilute sulphuric acid is 1.280 times as heavy as water. The case or container for the battery is usually made of hard rubber. Separators made of hard rubber or glass fibers are used to keep the plates apart.

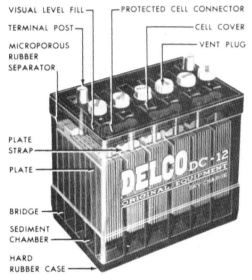

FIG. 10-21.   A typical lead-acid storage-battery construction. This is a 6-cell, 12-volt battery. (*Courtesy* Delco-Remy Division, General Motors Corporation)

A six-volt battery has three cells connected in series, positive to negative. Each cell has a voltage of about two volts. Six cells are necessary for a 12-volt battery. Each cell has a set of positive plates and a set of negative plates so that there is one negative plate on each side of each positive plate. There is always one more negative plate per cell than there are positive plates. All of the negative plates of one cell are connected together to form the negative terminal. The positive plates are connected to form the positive terminal. Each cell has enough electrolyte to cover the plates and enough separators to keep the plates from coming in contact with one another.

Batteries will discharge when being used. A chemical reaction takes place that changes the lead peroxide, sponge lead, and sulphuric acid to lead sulphate and water. The discharge equation is as follows:

$$PbO_2 + Pb + 2H_2SO_4 \rightarrow 2PbSO_4 + H_2O$$
lead peroxide + lead + sulphuric acid →
lead sulphate + water

Some hydrogen is usually given off in the process, but this is not shown in the above equation. This chemical reaction can be reversed by passing a direct current of proper voltage through the battery to recharge it.

Batteries that are partially or completely discharged will freeze in cold weather because the sulphuric acid becomes more dilute. Table 10-1 shows the relationship between the condition of the battery and its freezing point.

Care of Batteries

A well-serviced battery that is kept fully charged will give several years of reliable service.

Batteries should be kept fully charged. Recharge them when the specific gravity of the electrolyte reads below 1.200. The electrolyte should always cover the plates. Add distilled water whenever the electrolyte becomes too low. Keep the terminals and the connections clean.

Temperature has a great influence on the capacity of a lead-acid battery.

As Figure 10-22 shows at −20°F a battery is about 30 percent efficient, while at 80°F the battery is nearly 100 percent efficient. Also, if we assume that the power required to crank an engine at 80°F is 100 percent (Figure 10-22), then the power required to crank it at −20°F is 350 percent or three and one-half times greater. This helps explain why engines may be difficult to start in cold weather.

The tractor storage battery must be handled with care in order to avoid accidents. When the battery is being charged, some hydrogen is given off in the reaction. This is highly explosive. Do not light a match or use any other open flame near a battery. Do not disconnect a battery charger when the charger is operating. This may cause a spark which may ignite the hydrogen from the bat-

TABLE 10-1. BATTERY CONDITION AND FREEZING POINT

Percent Charged	Freezing Point °F	Freezing Point °C
100	−90	−68
75	−62	−52
50	−16	−26
25	−4	−20
discharged	+19	− 7

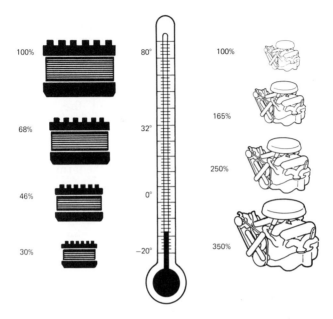

100%	80°	100%
68%	32°	165%
46%	0°	250%
30%	−20°	350%

FIG. 10-22.   How cold weather affects the battery and engine when starting. (*Courtesy* Deere and Company)

tery. Turn off the power to the charger before disconnecting it. When handling a battery, keep it away from your clothes. Battery acid is very detrimental to fabrics and will burn holes wherever it comes in contact with clothing.

IGNITION TIMING

The spark at the spark plug must occur at the right time to give good engine performance. When the engine is being started, the spark should occur when the piston is at the top dead center. However, when the engine is running, the spark must occur earlier because it takes a little time for the fuel charge to burn. It should be ready to exert its greatest force on the piston when the piston is ready to go down on the power stroke. When the engine is running, the spark at the plug is timed to occur about 25° before the piston reaches top dead center on the compression stroke. The exact number of degrees varies with engines and engine speed. High-speed engines usually have earlier spark timing than do slow-speed engines. Combustion-chamber design, compression ratio, and type of fuel will also affect engine-spark timing.

Tractor engines usually have marks on the flywheel or on the fan-drive pulley to show when the spark is to occur. When the marks are properly lined up, No. 1 piston will be at top dead center at the end of the compression stroke. Rotate the distributor of a battery-ignition system or adjust the magneto of a magneto-ignition system so that a spark will occur at No. 1 spark plug. This process is outlined in detail under Shop Projects.

FIG. 10-23. Timing an engine with battery ignition by rotating the distributor housing in the opposite direction to rotor travel, until a spark occurs at the number 1 wire. The engine must have the number 1 piston at TDC on the compression stroke. Tighten the lock screw on the housing when the spark occurs at number 1. (*Courtesy* American Oil Company)

SUMMARY

The battery-ignition system is an electrical device that increases the voltage of 6- or 12-volt batteries to 15,000 or 20,000 volts and delivers this high-voltage current to the spark plug at the right time to ignite the fuel charge.

The magneto-ignition system performs all of the functions that the battery system does and, in addition, it generates the electric current to produce the spark. Magneto- and battery-ignition systems have many parts in common. They function on some of the same basic principles. Battery systems are used on practically all newer tractors, but many old tractors are equipped with magneto systems. Single cylinder engines usually use magnetos.

Shop Projects:
Servicing the Ignition System

This covers several excellent projects that will extend the knowledge acquired in this chapter. Ignition systems will give reliable service for many years with only minor maintenance and repairs. These jobs can be done on the farm with only a few special tools.

A. Cleaning and adjusting the breaker points

1. Remove the distributor cap.

2. Examine the breaker points for pits and corrosion. File the points smooth with a point file. (Badly pitted or worn points should be replaced. If the points continue to pit rapidly, the condenser may also need replacing.)

3. Determine the proper breaker-point gap. See the operator's manual.

4. Turn the engine over until the cam holds the points in wide-open position.

5. Loosen the lock screw on the adjustable breaker point.

6. With a screwdriver, set the movable point so that the proper gap is obtained (Figure 10-8).

7. Turn the engine to the next high point on the cam and check the gap again. It should be the same as before.

8. Wipe the distributor cap on the inside with a clean, dry cloth and replace it on the distributor.

9. The cam angle may be checked with a cam angle meter.

B. Timing the battery-ignition system
    Each make of engine has a slightly different ignition-timing procedure. It is best to refer to the operator's manual furnished with the tractor to obtain specific details for a given engine. The general rules for timing an engine are as follows:

    1. Disconnect all spark-plug wires and remove the ignition unit from the tractor engine.

    2. Turn the engine with a hand crank until No. 1 piston is at top dead center on the end of the compression stroke. This can be determined in one of several ways. Use a, b, or c, whichever method is best suited to the engine.

        (a) On valve-in-head engines, it is easy to locate TDC on No. 1 piston by removing the valve cover and noting the opening and closing of the valves on cylinder No. 4 of a four-cylinder engine or on cylinder No. 6 of a six-cylinder engine. (Remember that pistons 1 and 4 travel together on a four-cylinder engine and 1 and 6 on a six-cylinder engine.) When No. 1 is coming up on compression, No. 4 (or 6) is coming up on exhaust. Therefore, when the exhaust valve just closes on No. 4 (or 6) and the intake valve is just ready to open, No. 1 is at TDC on the end of the compression stroke.

        (b) Remove the spark plug on No. 1 cylinder and with a stiff wire feel the position of the piston. Air will rush out of the spark-plug hole when the piston is on the compression stroke. Stop when the piston reaches TDC.

        (c) If a tractor engine has timing marks on the flywheel or fan-

FIG. 10-24.   Timing marks on the engine flywheel. It is usually necessary to remove a cover to find these marks. (*Courtesy* Massey-Ferguson, Limited)

drive pully (Figure 10-24), line up the marks according to the operator's manual.

3.   Remove the distributor cap and turn the ignition unit by hand until the rotor in the distributor points to No. 1 terminal and the breaker points are just opening.

4.   Keep the ignition unit shaft in this position and mount the unit on its drive mechanism on the engine.

5.   Connect the spark-plug wires, starting with No. 1 and continuing in the direction of rotor rotation and in sequence of the firing order of the engine.

6.   The engine should now be in time, but may need some slight adjustment to put it in exact time with the marks on the flywheel.

7. Check the timing as follows:

    (a) Crank the engine until the timing marks are in line.

    (b) Remove the distributor cap and pull the center wire out of its socket in the distributor cap. Hold the end of this wire about one-eighth inch from the engine block.

    (c) Loosen the distributor clamping screw and rotate the entire distributor housing in the direction of the rotation of the cam until the breaker points are closed. This usually requires less than one-eighth of a turn. (Figure 10-23)

    (d) Turn on the ignition switch.

    (e) Now turn the distributor *slowly* the opposite direction from above until the breaker points open and a spark jumps across the gap between the end of the wire and the engine block.

    (f) Tighten the distributor-housing clamp screw and replace the cap and the wire. The engine is now in time and should give satisfactory performance.

C. Timing the magneto-ignition system

This job is similar to that of timing the battery-ignition system. Proceed as follows:

1. See Step 1 under Timing the battery-ignition system.

2. See Step 2 under Timing the battery-ignition system.

3. Remove the distributor cap and turn the magneto by hand until the rotor in the distributor housing points to the No. 1 terminal (usually marked) on the distributor cap.

4. Keep the magneto shaft in this position and mount the magneto in its place on the engine.

5. Connect the spark-plug wires, starting with No. 1 and continuing in the direction of rotor rotation and in sequence of the firing order of the engine.

6. The engine is now approximately in time but will need a final checking and adjustment as follows:

Turn the engine over until No. 1 piston is close to the firing position. Note the top dead-center mark on the flywheel. The impulse coupling should trip when the TDC mark for No. 1 piston reaches its corresponding mark on the flywheel housing (Figure 10-24). If the impulse is late, loosen the magneto mounting and rotate the magneto in a direction opposite to the direction of rotation of the magneto shaft. If the impulse is early, rotate the magneto in the direction of the magneto-shaft rotation. When the engine has been properly timed so that the impulse trips at TDC, tighten the magneto mounting bolts.

D. Cleaning the battery

The area around the terminals of storage batteries becomes corroded after some use. It is a simple matter to remove this corrosion with baking soda and water.

1. Sprinkle a liberal amount of baking soda around the corroded terminals.

2. Pour on a little water and stir. Allow the baking soda and water to dissolve the corrosion completely.

3. Flush the battery top thoroughly with water to remove all soda and corrosion.

4. Grease the battery terminals with chassis lubricant or similar grease. The grease will keep corrosion from forming on the terminals for many months.

Questions

1. What voltage is necessary to make electricity jump the gap at the spark plug? Is this affected by the compression ratio?

2. How is this voltage produced in the battery-ignition system? In the magneto-ignition system?

3. What parts used in the battery-ignition system are also used in the magneto system?

4. List the parts that are in the primary circuit of an ignition system. List the parts that are in the secondary circuit.

5. Why must the spark be timed to occur before the piston reaches TDC

on the compression stroke? How many degrees before TDC should the spark occur?

6. What service jobs on ignition systems can normally be done on the farm?

7. Why are spark plugs made in several heat ranges? Under what conditions should a cold spark plug be used in an engine? When should a hot spark plug be used?

8. Why is a flat feeler gauge unsuited for measuring the spark-plug gap?

9. Will a fully charged lead-acid battery have more resistance to freezing than one that is partly charged? Why?

10. How does the temperature of the storage battery affect its capacity?

11. How can corroded battery terminals be cleaned? Why should cleaned terminals be coated with grease?

12. What safety precautions should be followed in handling or charging storage batteries?

13. Explain the procedure to use in timing a battery-ignition system so that the spark will occur at the right time.

14. Explain the methods of determining when No. 1 piston of a four-cycle engine is at the top dead center on the compression stroke.

References

Fundamentals of Service, Engines, 3rd Edition, Deere & Company, Moline, Illinois, 1977.

Fundamentals of Machine Operation, Preventive Maintenance, Deere & Company, Moline, Illinois, 1973.

Stone, A. A. and H. E. Gulvin, *Machines for Power Farming,* John Wiley and Sons, Inc., New York, New York, 1967.

The Tractor Electrical System, American Association for Vocational Instructional Materials, Athens, Georgia,

Tractor Maintenance, American Association for Vocational Instructional Materials, Athens, Georgia, 1975.

11 Electrical Accessories

Tractors may be equipped with many electrical accessories. These include battery charging equipment, starters, lights, various gauges, radios, tape players, air conditioning controls, and others. We will discuss battery charging equipment, starters, and lights.

BATTERY CHARGING EQUIPMENT

Two types of battery charging equipment are in general use: the generator and alternator.

The Generator

The generator has been used for many years and will be found on many older tractor engines. Its purpose is to keep the battery charged to provide electric power for the ignition system and electrical accessories. Alternating current is generated by this unit. However, the action of the commutator converts the alternating current to direct current.

The generator on the farm tractor operates on the principle of electromagnetic induction, as described under the ignition system. Figure 11-1 shows the generator with its pole shoes (which are actually electromagnets), commutator, brushes, and field coil. As the armature of the generator is rotated between the electromagnets, the windings cut lines of force and a current is induced in the charging circuit.

A generator must have an automatic switch to disconnect it from the battery when it is not turning or when it is turning at slow speed. This automatic device is called a cutout (Figure 11-2). As the generator starts turning, a small current is generated through the grounded shunt winding. This energizes the soft iron core of the cutout and closes the points of the cutout. When the points are closed, the charging circuit is completed and the current from the generator will go to the battery. When the engine is running slowly, the current is not sufficient to close the points and no charging takes place. When the engine stops, the cutout points open and prevent the battery from being discharged through the generator.

The generator is equipped with a current and voltage regulator that automatically balances the generator output with the needs of the battery (Figure 11-3).

The regulator shown in Figure 11-3 contains three coils. The coil on the left is a cutout relay which is also shown in Figure 11-2. It is usually combined with the other units shown to form the regulator. The center coil regulates the current (ampere) output of the generator. It is wound with wire from the generator

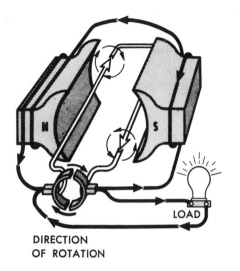

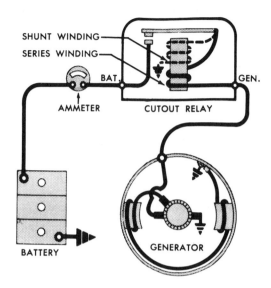

FIG. 11-1.   Generator diagram showing the field circuit around poles N and S. The armature rotates between the poles. Current for the load is picked up from the commutator by the brushes. Arrows indicate the direction of the current flow. (*Courtesy* Delco-Remy Division, General Motors Corporation)

FIG. 11-2.   The cutout relay is on the automatic switch that disconnects the generator from the battery when the generator is not charging. The points in the relay close when charging begins and open when the engine stops and the generator no longer charges. (*Courtesy* Delco-Remy Division, General Motors Corporation)

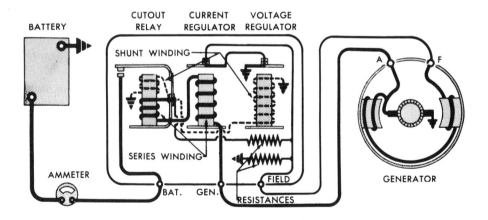

FIG. 11-3.   A regulator that controls the generator output to meet the needs of the battery. It controls voltage and current, and also has a cutout. (*Courtesy* Delco-Remy Division, General Motors Corporation)

charging circuit. When the current (ampere output) becomes too great, the points in the field circuit above this coil will be opened, thus causing the field circuit to be grounded through the resistance unit (lower right). The increased resistance in the field circuit will stabilize the output of the generator. The coil on the right regulates the voltage of the charging circuit. When the voltage becomes too high the points above the voltage coil will open. This action also causes the field circuit to be grounded through the resistance unit (lower right). This stabilizes the generator voltage. Regulators can be adjusted for maximum current and voltage.

When the battery is in a discharged or partly discharged condition, the current output of the generator tends to be high, but it is controlled by the current regulator. When the battery is charged, the voltage regulator will prevent excessive voltage buildup.

The ammeter (Figure 11-4) in the electrical circuit is a measuring instrument which indicates the current flow from the generator to the battery and from the battery to the ignition, lights, and other accessories. The flow from the generator to the battery will show on the charge side of the ammeter. The current being used by the ignition system, the lights, and the other electrical accessories on the tractor will be indicated on the discharge side when the engine is not running. The ammeter may also serve as an indicator to show whether the generator is functioning properly, and whether there are short circuits in the system. An extremely heavy discharge rate would indicate a short circuit. The failure of the ammeter to indicate a charge would show that the generator may not be functioning properly.

FIG. 11-4. An ammeter indicates the current flow from the generator to the battery on the charge or + side. Flow from the battery is shown on the discharge or − side. (Photo by Authors)

Servicing the Generator

The generator will give many hours of trouble-free performance with a minimum of maintenance. Occasionally it may be necessary to clean the commutator (Figure 11-5) with #00 (fine) sandpaper. Hold the sandpaper tightly against the commutator segments while the engine is running slowly. This will remove corrosion and any slight irregularities that may occur on the commutator after many hours of use. Emery cloth should not be used for this purpose, because it is a conductor of electricity and might short the commutator segments. The bearings of the generator should be oiled with a few drops of light oil several times during the season. Oil should be used sparingly, because excessive amounts may get into the generator windings and cause damage.

FIG. 11-5.   The generator or commutator can be cleaned with fine sandpaper. Hold the sandpaper against the commutator while the engine is running slowly. Do not sand more than is necessary to remove the corrosion. (Photo by Authors)

FIG. 11-6.   Place a few drops of oil in the generator bearings several times during the season. Use a light oil such as SAE #10. (Photo by Authors)

THE ALTERNATOR

The alternator is similar in principle to the generator. In the generator the field circuit is stationary. In Figure 11-1 the field circuit is wound around the poles N and S. The armature revolves within this field. It generates alternating current, but the commutator converts it to direct current. In the alternator (Figure 11-7) the field coils, including the bar magnet, rotate. As this field rotates, it cuts across the stationary outer wire loop, inducing a voltage in this loop. Again an alternating current is generated. However, in the alternator the alternating current is converted to direct current through a system of diodes. Item number 3 in Figure 11-8 shows the location of the diode plate assembly. Diodes are electronic devices that will permit the flow of electric current in one direction only.

The rotating field circuit of the alternator is called the rotor (Figure 11-8). The stationary loop is known as the stator. The diodes serve as a rectifier. They convert alternating current to direct current. Thus, the alternator, when equipped with diodes, becomes a direct current generator.

The alternator has two advantages over the generator: (1) it does not have a commutator to cause brush and cleaning problems and (2) it can.be turned faster than the generator, thus increasing its output for slow engine speeds.

On a tractor equipped with an alternator, it is best to consult the operator's manual for instructions on installing and servicing the alternator and the battery.

THE STARTER

The starting motor, or starter, is a series wound electric motor that is mounted near the engine flywheel so that it can engage the flywheel gear to start the engine. It is usually a 12 volt system, although some large diesel engines may have a 24 volt system.

When the starter switch or key

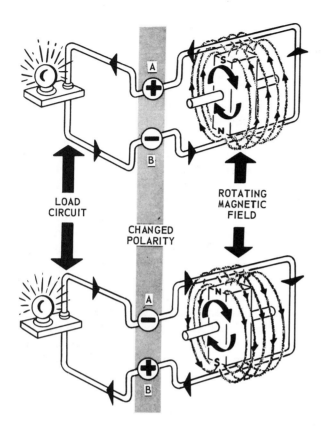

LOAD
CIRCUIT

CHANGED
POLARITY

ROTATING
MAGNETIC
FIELD

FIG. 11-7. The principle of the alternator. Note the changed polarity top and bottom. It generates alternating current. (*Courtesy* Deere and Company)

switch is turned "on" the solenoid is activated and it will engage the starter gear with the flywheel gear (Figures 11-9 and 11-10). The key switch carries only the relatively low current necessary to activate the solenoid. The heavy starting current required to crank the engine bypasses the key switch (Figures 11-10 and 11-11).

After the starter gear has engaged the flywheel gear, the starting motor will crank the engine (Figure 11-11). When the engine starts the two gears will be disengaged.

FIG. 11-8. Cutaway view of an alternator showing (1) rotor, (2) stator, (3) diode plate assembly, (4) front housing, (5) fan and pulley assembly, and (6) rear housing. (*Courtesy* Autolite-Ford Parts Division, Ford Motor Company)

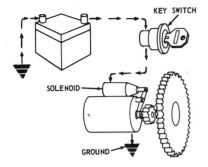

FIG. 11-9. With the starter switch turned "on," the starter gear is ready to engage the large flywheel gear. (*Courtesy* Deere and Company)

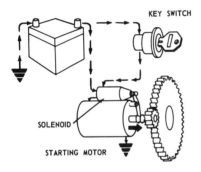

FIG. 11-10. The starter gear has engaged the flywheel gear. (*Courtesy* Deere and Company)

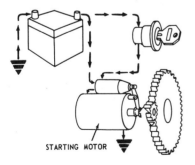

FIG. 11-11. The starter is cranking the engine. When the engine starts, the starter gear will be disengaged from the flywheel gear. (*Courtesy* Deere and Company)

The starting motor is powerful and will be a heavy drain on the battery while starting the engine. The engine-starting period is usually short so that the generator can recharge the battery in a short time. If the starting is prolonged, the drain on the battery may be extremely heavy and recharging would take considerable time. Cold-weather starting requires a powerful, well-maintained starting unit and a fully charged battery in good condition.

Figure 11-12 shows a starter that has a "shift lever" to engage the starter gear with the flywheel gear. The same lever also operates the starter switch. When the engine starts, the shift lever is released, the gears disengage, and the starter switch is turned off. Figure 11-13 shows a starter with a solenoid that will engage the pinion gear with the flywheel gear when the starting switch is closed. The solenoid also engages the starting

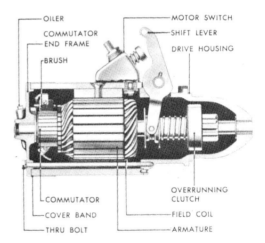

FIG. 11-12. A typical starting motor. It is mounted on the tractor with the starter drive gear in position to engage the gear around the tractor flywheel when the switch is closed. When the switch is open, the starter gear is released from the flywheel gear. (*Courtesy* Delco-Remy Division, General Motors Corporation)

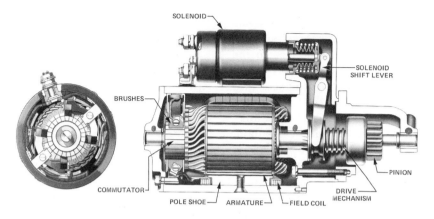

FIG. 11-13. Starter with solenoid to engage pinion with gear on flywheel. (*Courtesy* Delco-Remy Division, General Motors Corporation)

motor and causes it to crank the engine. When the engine starts the pinion gear is released and the solenoid and gear return to their original position.

The connections from the battery to the starting motor and the ground connection from the battery must be kept clean and tight. The commutator on the starter can be cleaned with sandpaper in a manner similar to that used in cleaning the generator commutator.

TRACTOR LIGHTS

The lighting circuit is a system of wires that permits the lights to draw current from either the battery or the generator when the light switch is on. A typical lighting circuit is shown in Figure 11-14. Electricity must have a complete path or circuit from the battery through the lights and the switch, and back through the battery. Tractors are gen-

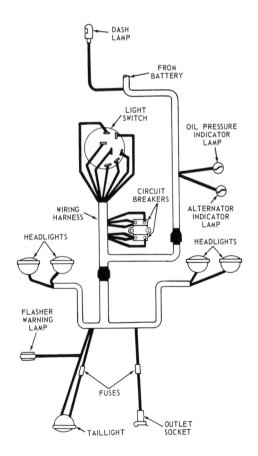

FIG. 11-14. Schematic diagram of a typical tractor lighting circuit. (*Courtesy* Deere and Company)

erally equipped with sealed-beam lights similar to those used in automobiles.

In order to prevent damage to the circuit when an overload or short circuit occurs, safety devices are provided. Two types are in general use: the fuse and the circuit breaker. The fuse is a "weak link" in the system that will fail before damage is done to the circuit. A burned out fuse should be replaced with one of the same ampere rating. The circuit breaker is an automatic switch that will go to the "off" position when the circuit is overloaded. When the overload condition is corrected and the circuit breaker has had a chance to cool, it can again be switched to the "on" position for further use.

FIG. 11-15. Adjust generator or alternator drive belt tension according to manufacturer's recommendation. (*Courtesy* Deere and Company)

Generator and alternator drive belts must be kept tight in order to get optimum performance from these units. Belts that are too loose will slip and cause low current output and rapid belt wear. The belt tension can be adjusted as shown in Figure 11-15. Follow the manufacturer's recommendation for proper belt tension.

SUMMARY

The generator or alternator, starter, and lights are standard equipment on most farm tractors. Generators, alternators, and starters operate on the principle of electromagnetic induction.

Generators and alternators supply current to the battery. Generators have a cutout to prevent the battery from discharging when the engine is not running or running at low speed. Voltage and current controls are also used on tractor generators. These balance the generator output to the needs of the electrical system.

Alternators serve the same purpose as generators, but have the advantages of greater output at low engine speeds and less maintenance. They use diodes to prevent the battery from being discharged when the engine is not running.

The starter is used to crank the engine and will be a heavy drain on the battery, especially during cold weather. A well-maintained starting motor is essential.

The lighting system is a simple electric circuit that makes the electric energy of the battery available at the lights.

All of these units require little attention but should be serviced as needed. They will then give many years of service and will often outlast the tractor. It is best to follow closely the recommendations listed in the operator's manual.

Shop Project:
Generator Care

It is best to refer to the owner's manual for proper charging rates of the generator and other adjustments. If the ammeter does not show a proper charging rate, make adjustments according to the owner's manual. The adjustment will depend upon the type of generator and control system used on the tractor. Check other generator adjustments as follows:

A. The slack in the drive belt should be approximately one-quarter inch. A drive belt that is too loose will result in slippage and a low charging rate. If the drive belt is too tight, it will cause excessive wear on the bearings.

B. Remove the cover band near one end of the generator housing and examine the commutator. If the commutator segments are rough and corroded (usually indicated by a black appearance), it will be necessary to remove the corrosion with #00 (fine) sandpaper held lightly against the commutator (Figure 11-5). Run the engine slowly while cleaning the commutator segments. If the commutator is extremely rough and if the insulators between the commutator segments are protruding above the commutator segments, it will be necessary to remove the generator and take it to a service shop for reconditioning.

C. The cutout and the voltage regulator are intricate mechanisms that require careful adjustments. It is best to perform these adjustments in a shop equipped with testing apparatus for this purpose. A student should see that all connections to the cutout and the voltage regulator are clean and that the unit itself is free of dust.

D. Lubricate the generator bearings with a few drops of light oil if an oil cup is provided. This should be done when the engine is not running. The oil cups of the generator bearings are found at each end of the generator. Each one is equipped with a small hinged cap to keep out dust. This cap must be raised in order to get oil to the generator bearings (Figure 11-6).

Questions

1. What is the purpose of the generator or alternator on a tractor?

2. How can the charging rate of a generator be controlled? Explain two methods.

3. What is the mechanism that is used to keep the generator from discharging the battery when the engine is not running? Explain the operation of this unit.

4. What is the purpose of the ammeter in the charging circuit?

5. In outward appearance the generator and the starter are quite similar. How do they differ in internal windings?

6. What service jobs should be performed regularly on the generator? On the starter?

7. What is the purpose of a fuse in the lighting circuit of the tractor?

8. On the belt-driven generators, what damage might result if the belt is too loose? Too tight? How should the belt be adjusted for proper tension?

9. What are the advantages of alternators compared to generators?

10. What device keeps the alternator from discharging the battery when the engine is not running?

References

Autolite Alternator Service and Testing Procedures, Ford Motor Company, Dearborn, Michigan, 1966.

Fundamentals of Service, Electrical Systems, Deere & Company, Moline, Illinois, 1972.

Fundamentals of Service, Engines, Deere & Company, Moline, Illinois, 1977.

Jones, Fred R., *Farm Gas Engines and Tractors*, McGraw-Hill Book Company, New York, New York, 1963.

The Tractor Electrical System, American Association for Vocational Instructional Materials, Athens, Georgia.

12   Diesel Engines

Modern high-speed diesel engines such as those used in farm tractors are very similar in outward appearance to gasoline and LPG spark-ignition engines. However, diesel engines differ in these ways:

1. The fuel charge is injected into the combustion chamber under high pressure. (In gasoline and LPG engines the fuel and air are mixed at the carburetor.)
2. The injected fuel charge is ignited by the high temperature of the highly compressed air in the combustion chamber. This requires a compression ratio of about 15 to 1 or higher. No electric spark is necessary for ignition.
3. The air entering the cylinders is not throttled. The engine usually operates with an excess of air.

   Diesel engines can be designed to operate either as two-stroke cycle engines or four-stroke cycle engines. Both are used on farm tractors.

THE TWO-STROKE CYCLE DIESEL ENGINE

The construction, operation, and efficiency of the two-cycle diesel engine differ from those of the gasoline-burning two-cycle engine. As the piston nears the end of the first downward stroke, the exhaust valve opens and the air-intake ports are uncovered. Unthrottled air is blown through the ports and out the exhaust valve, which results in a complete clearing of the combustion space as well as the filling of the chamber with fresh air. As the piston moves upward, the intake ports are covered by the piston and the air is compressed to a high degree, causing it to become very hot, approximately 1100° F (593.3° C). Fuel under high pressure is forced into the cylinder as the piston nears the top of the compression stroke.

The burning of fuel mixed with the hot air forces the piston downward on its power stroke. Figure 12-1 shows the four events that make up the cycle of this engine. Notice that only two strokes of the piston, compression and power, are required.

Exhaust and fresh-air intake take place at the end of the power stroke and before the compression stroke starts. High engine efficiency is obtained by the complete clearing of exhaust gases from the combustion chamber, a thorough filling of the chamber with clean air, and the injecting of fuel into highly compressed air, near the end of the compression stroke.

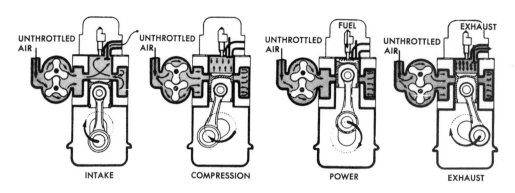

FIG. 12-1.   Operating principles of the two-cycle diesel engine. (*Courtesy* Allis-Chalmers Manufacturing Company)

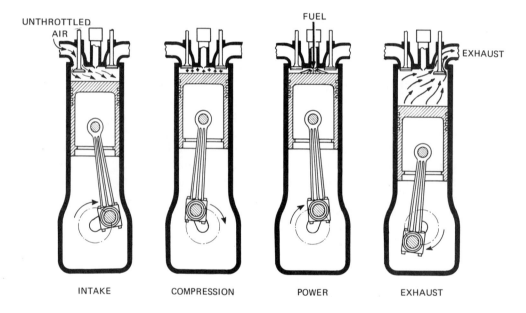

FIG. 12-2.   Four-stroke cycle diesel-engine operating principles. Fuel is injected near the end of the compression stroke. (*Courtesy* American Oil Company)

THE FOUR-STROKE CYCLE DIESEL ENGINE

The basic principles of the four-stroke cycle diesel engine and four-stroke cycle gasoline engine are similar except that:

1. The diesel takes in air only on the intake stroke.
2. The fuel charge for the diesel is injected under high pressure near the end of the compression stroke.

The diesel engine needs no electrical ignition system. The fuel charge is ignited as it is injected into the hot, 1100° F (593° C), highly compressed air in the combustion chamber. The compression ratio of the diesel engine is about 15 to 1. This is necessary to develop temperatures needed for self-ignition. Fuel injection begins at about 20° before TDC on the compression stroke and continues for a number of degrees of crankshaft travel. Under light loads the cutoff point will come earlier than under heavy loads. The governor regulates the length of the injection period to fit the load demands on the engine.

DIESEL-ENGINE CONSTRUCTION

Because of the higher pressures present in the diesel engine, its construction must be heavier than that of a gasoline engine. Diesels usually have heavier piston pins, connecting rods, bearings, and crankshafts than those found in gasoline engines.

DIESEL COMBUSTION CHAMBERS

When diesel fuel is injected into the combustion chamber, a certain amount of delay is experienced in the beginning of combustion. This delay can be influenced by the type of fuel being used and by the combustion-chamber design. Figure 12-3 shows the types of combustion chambers commonly used on diesel engines.

The open-chamber or direct-injection method usually results in a greater ignition delay because more fuel needs to be injected before a combustible mixture is formed. However, the direct-injection chamber is usually designed to give the air and fuel a swirling action to insure good combustion.

The precombustion and turbulence chambers normally result in less delay because only a part of the air used is in the precombustion or turbulence chamber, and less fuel is required to provide a combustible mixture for the beginning of ignition. As the fuel burns, the flame spreads to the main combustion chamber. Also, the turbulence chamber, because of its design, gives the fuel charge a swirling action which results in good

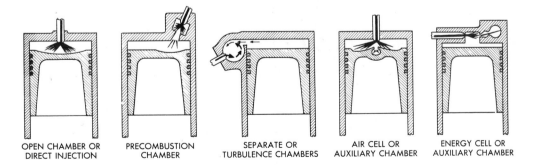

OPEN CHAMBER OR
DIRECT INJECTION

PRECOMBUSTION
CHAMBER

SEPARATE OR
TURBULENCE CHAMBERS

AIR CELL OR
AUXILIARY CHAMBER

ENERGY CELL OR
AUXILIARY CHAMBER

FIG. 12-3. Typical diesel combustion-chamber designs. (*Courtesy* American Oil Company)

mixing of air and fuel and good combustion in the main chamber.

The auxiliary chamber (air cell and energy cell) is a small cell so located that fuel from the injection nozzle will be directed into it. The fuel in the cell burns and causes a turbulence as it spreads to the main combustion chamber. This results in good mixing of air and fuel and good combustion. The starting characteristics of a diesel engine using this type of combustion chamber are usually quite good.

THE FUEL-SUPPLY SYSTEM FOR DIESEL ENGINES

The diesel fuel supply system may vary somewhat from one engine to another. However, most systems include the parts shown in Figure 12-4. The fuel flows from the fuel tank to the transfer pump through the fuel filters, injection pump, and injection nozzles. The injection nozzles deliver the fuel to the combustion chamber. The fuel enters the combustion chamber in the form of a fine spray. Pumps and injectors are designed to provide some "leakage" for lubrication purposes. This fuel is returned to the fuel tank through the fuel return line.

The diesel fuel supply system meters the fuel to the engine. Each injection of fuel is quite small. The amount of each injection is accurately controlled by the metering system of the injection pump. Some idea of the injection pump's job can be gained by calculating the number of fuel injections that are required by an engine in a given time. If a six cylinder four stroke cycle diesel engine running at 2200 RPM uses five gallons (18.93L) of diesel fuel per hour, then the injection pump system must divide

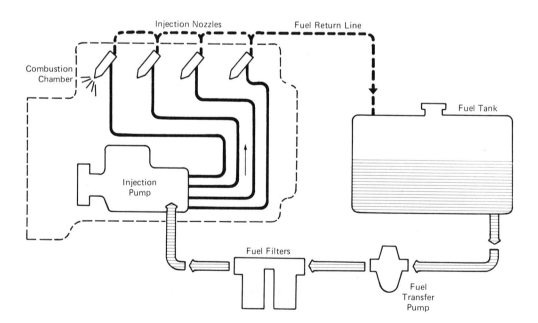

FIG. 12-4.   A diesel fuel supply system. (*Courtesy* Deere and Company)

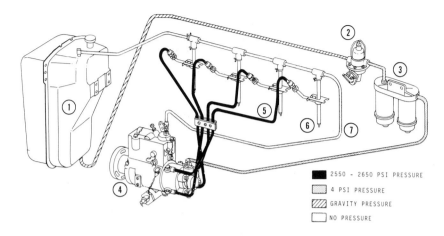

2550 - 2650 PSI PRESSURE

4 PSI PRESSURE

GRAVITY PRESSURE

NO PRESSURE

FIG. 12-5.   A diesel fuel supply system. The fuel flow is from tank (1) to the transfer pump (2), to the filters (3), then to the high-pressure pump (4) through the high-pressure lines (5) to the injection nozzles (6). Excess fuel is returned to the tank through the return line (7). (*Courtesy* Deere and Company)

this amount of fuel into 396,000 parts in one hour. Also, it must adjust the size of the injections to meet the fuel requirements of the engine.

The diesel fuel system operates under two pressure levels. The low pressure portion of the system includes the fuel tank, transfer pump, and fuel filters. The high pressure portion of the system includes the injection (high pressure) pump, the high pressure lines leading to the injection nozzles, and injection nozzles.

Excess fuel is returned to the fuel tank under nearly zero pressure through the fuel return line. Figure 12-5 shows a system for a four cylinder diesel engine.

Transfer Pumps

Fuel transfer pumps are used to transfer fuel from the fuel tank through the fuel filters and to the high pressure injection pump. (On some engines a gravity system is used to supply the fuel to the injection pump.) The transfer

pump may be of the diaphragm type shown in Figure 12-6. The rocker arm is driven from a cam on the engine. The rocker arm operates the diaphragm which pumps the fuel through a valve

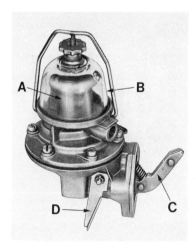

FIG. 12-6.   A diaphragm type of fuel transfer pump. Parts identified are: (A) fuel strainer, (B) fuel bowl, (C) rocker arm, and (D) hand primer lever. (*Courtesy* Deere and Company)

arrangement inside the pump housing.

Transfer pumps are also used within the high pressure pump housing. They deliver fuel from the supply line to the high pressure injection pumps. The injection pumps shown in Figure 12-8 and 12-14 both contain transfer pumps.

Fuel Filters

The injection pump and injectors on a diesel engine have very close-fitting parts. This is why diesel fuel and fuel systems must be kept clean. There are usually two filters (Figure 12-7) through which the fuel must pass on its way from the fuel tank to the injection pump.

Permitting dirt to enter the fuel system would cause the pump and injector parts to wear rapidly. These parts are lubricated by the diesel fuel.

Service of the diesel fuel system should be handled by qualified personnel. The operator's manual for an engine will specify what service is necessary

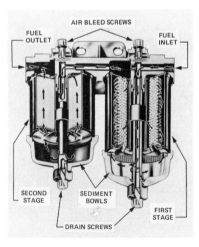

FIG. 12-7.   A series type fuel filter. All fuel passes through the right hand filter before it is filtered by the left hand filter. Arrows show direction of fuel flow. (*Courtesy* Deere and Company)

and at what intervals this servicing should be done. Since engines will vary, no attempt will be made here to outline the service adjustments.

FUEL INJECTION PUMPS

Two types of fuel injection pumps are generally used on farm engines. There is the in-line type which has an individual pumping element for each cylinder (Figure 12-8). The other is the distributor type which has one injection pump for all of the cylinders, but has a fuel distributor to take the fuel charge to the proper cylinder at the right time. Figure 12-14 shows a distributor pump.

In-line Injection Pump

The individual pumping elements in the in-line type pump are of the plunger type. Each plunger or pumping element is operated by the camshaft of the pump (Figure 12-8). The plungers always have the same length of stroke. The quantity of fuel supplied by each pumping unit to the cylinder is metered by the helix machined on the side of the plunger (Figure 12-9). When the plunger is at the bottom of its stroke, the cavity above the plunger and around the helix are filled with fuel. As the cam raises the plunger, the ports in the barrel are closed thus trapping the fuel in the area above the plunger and around the helix. The delivery of fuel through the injection nozzle and to the cylinder begins at this point. Delivery continues until the edge of the helix exposes the control port (at the right in Figure 12-9).

When the plunger is in the "no fuel delivery" position (Figure 12-10), the vertical slot in the plunger is exposed to the control port. Fuel can escape through

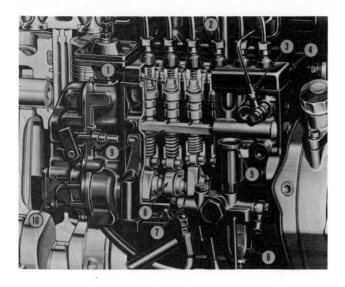

FIG. 12-8. In-line injection pump. One individual pumping element for each cylinder.

(1) Individual pumping element
(2) Injection line
(3) Leak-off line
(4) Pump housing
(5) Hand primer
(6) Sediment bowl
(7) Fuel transfer pump
(8) Camshaft
(9) Control rack
(10) Governor

(*Courtesy* Deere and Company)

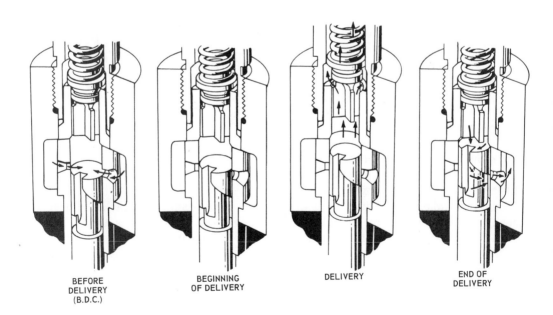

BEFORE
DELIVERY
(B.D.C.)

BEGINNING
OF DELIVERY

DELIVERY

END OF
DELIVERY

FIG. 12-9. A plunger operating at maximum delivery. (*Courtesy* Deere and Company)

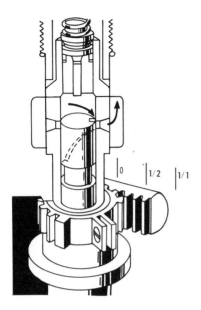

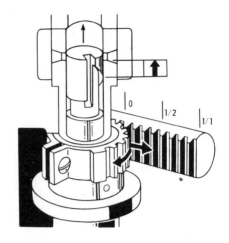

FIG. 12-11.   Plunger in partial fuel delivery position. (*Courtesy* Deere and Company)

FIG. 12-10.   Plunger in no fuel delivery position. (*Courtesy* Deere and Company)

the slot and port so no fuel delivery takes place in this position.

Partial load fuel delivery takes place between the no delivery and maximum delivery positions of the plunger (Figure 12-11). Notice that the plunger has been rotated by the governor so that the plunger travels through part of its stroke before the edge of the helix uncovers the control port.

Maximum fuel delivery takes place when the plunger is rotated by the governor to the position shown in Figure 12-12. In this position the plunger travels its maximum stroke before the edge of the helix uncovers the control port. Figure 12-13 shows another view of a diesel fuel metering device.

As mentioned before the governor rotates the plunger to provide the amount of fuel required to meet the load and speed conditions imposed upon the engine. Figure 12-12 and the two previ-

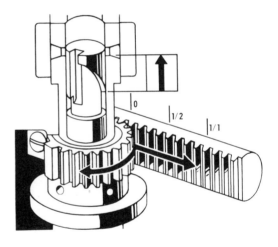

FIG. 12-12.   Plunger in maximum fuel delivery position. (*Courtesy* Deere and Company)

ous Figures show the rack and gear arrangement. The rack is the "straight" gear that rotates the plunger. The rack is moved back and forth by the governor. The location of the rack with respect to the other injection pump parts is shown in Figure 12-8. Figure 12-18 shows a schematic diagram of the governing system.

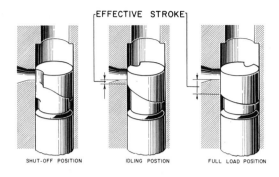

EFFECTIVE STROKE

SHUT-OFF POSITION   IDLING POSTION   FULL LOAD POSITION

FIG. 12-13. Helix-type metering device for diesel engines. (*Courtesy* Caterpillar)

Distributor Type Injection Pump

The distributor type fuel injection pumps shown in Figures 12-14 and 12-15

meter the amount of fuel that enters the high-pressure system of the pump. The amount of fuel entering the high-pressure system is controlled by the throttle valve which is linked to the governor. At light loads less fuel enters the high-pressure injection area. Thus the opposed plungers (Figure 12-15) of the pump do not return to the full length of the stroke. Therefore, the beginning of injection is retarded and less fuel will be injected. At full load, maximum fuel delivery is required. The plungers are forced all the way out and the space between them is filled with fuel. When the dual roller actuating cams contact the plungers, a full charge of fuel is delivered. The two opposed plungers are

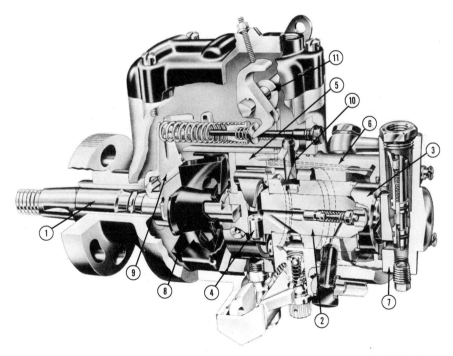

FIG. 12-14. A single-cylinder opposed plunger, inlet metering, diesel fuel-injection pump showing (1) drive shaft, (2) distributor rotor, (3) transfer pump, (4) pumping plungers (high-pressure pump), (5) internal cam ring, (6) hydraulic head, (7) end plate, (8) governor, (9) governor arm, (10) metering valve, and (11) shutoff lever. (*Courtesy* Standard Screw Company, Hartford Division)

DISCHARGE

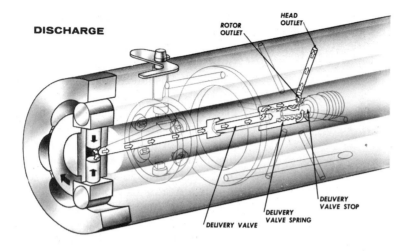

CHARGE

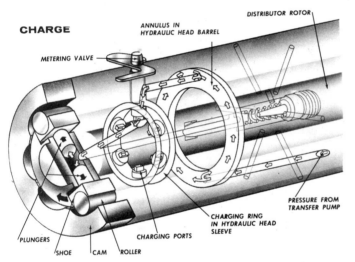

FIG. 12-15.   Fuel-distribution system of the pump shown in Fig. 12-14. (*Courtesy* Standard Screw Company, Hartford Division)

mounted in a rotating drum. This drum has a set of outlets for distributing fuel under high pressure to the combustion chamber of each engine cylinder. The number of dual cams necessary to actuate the plungers is equal to the number of cylinders in the engine.

The diesel fuel distributor delivers a fuel charge to the combustion chamber of the diesel engine in very much the same way as the distributor of a gasoline engine delivers an electrical spark to the spark plug and combustion chamber. The function of the fuel injection nozzles is to direct the metered quantities of fuel received from the fuel-injection pump into the engine combustion chambers in a definite spray pattern. The valve of each fuel-injection nozzle is operated hydraulically by the pressure of the fuel delivered by the pump.

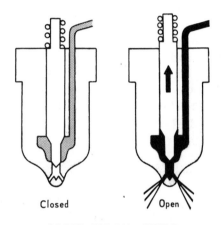

INWARD–OPENING NOZZLE

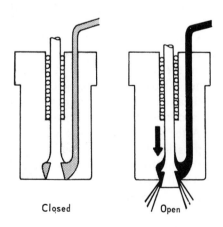

OUTWARD–OPENING NOZZLE

FIG. 12-16. Two types of injection nozzles used on diesel engines. (*Courtesy* Deere and Company)

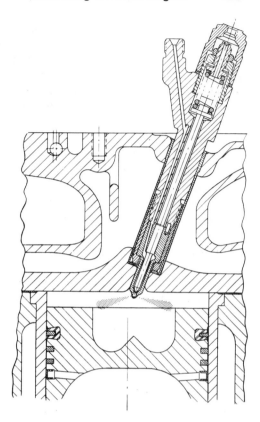

FIG. 12-17. Diesel fuel-injection nozzle shown in relation to the piston and the combustion chamber. (*Courtesy* Allis-Chalmers Manufacturing Company)

GOVERNING A DIESEL ENGINE

Diesel fuel is injected under high pressure. The fuel-metering device on the injection-pump plunger has a helix (inclined slot in Figure 12-13). As this plunger is rotated by the governor, clockwise or counterclockwise, it meters more or less fuel to the injection nozzles because the effective stroke of the fuel-injection pump plunger is varied as the plunger is rotated (Figure 12-18). This is accomplished by the helix on each plunger which allows fuel above the plunger to flow back through the inlet port whenever it is uncovered.

In the distributor type injection pump shown in Figure 12-14, the amount of fuel entering the high pressure system is controlled by the governor. The fuel passes through a governor-controlled metering valve which supplies the correct amount of fuel to the high pressure pump to accommodate the fuel requirements of the engine. The metering valve is controlled by a centrifugal governor. In most governors the position of the speed control lever (throttle) determines the tension in the governor spring. As engine speed increases, the governor weights will move outward due to the centrifugal force of rotating weights. When the centrifugal force of the weights balances the tension in the governor spring, the engine will be governed at that throttle or speed setting.

As the load on the engine increases, the engine speed will tend to drop. The governor weights will move together. This action of the weights causes more

fuel to be supplied to the engine and governed speed will be maintained. As the load on the engine becomes less, engine speed tends to increase and the governor weights will move apart. This action of the weights causes less fuel to be supplied to the engine and the engine will continue at the governed speed. A sensitive governor will maintain a uniform engine speed with only slight variations due to changing loads on the engine.

In the diesel governor, as in the gasoline and LPG engine governors, the speed control lever on the driver's platform adjusts the tension in the governor spring. This governor also functions at all engine speeds.

DIESEL STARTING SYSTEMS

Some diesels use an auxiliary gasoline engine for starting. This is a very dependable system, but is more costly than the electric system. The small engine is attached to the diesel by means of a clutch. The gasoline engine is used to crank the diesel until the diesel starts. Exhaust from the starting engine is used to heat the intake manifold of the diesel engine. Also, the compression on the diesel can be released temporarily to aid the gasoline engine in motoring the diesel. After a brief warm-up period, the compression lever is returned to normal and the throttle valve opened so that the diesel will start.

Most farm diesels use electric starting motors similar to those used on gasoline tractors. However, they are heavy-duty units. They are operated from the battery on the tractor. Starting must be prompt in order to conserve battery energy. In warm weather, starting is

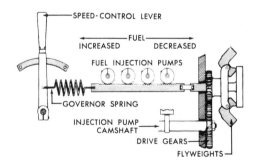

FIG. 12-18. The governor flyweights move the geared rack that rotates the fuel-injection pump metering helixes (Figure 12-13) to supply the correct amount of fuel to the engine. Moving the speed-control lever adjusts the tension in the governor spring. This system governs at all engine speeds. (*Courtesy* American Oil Company)

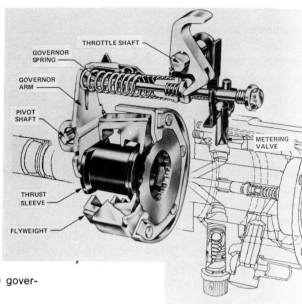

GOVERNOR
SPRING

THROTTLE SHAFT

GOVERNOR
ARM

PIVOT
SHAFT

METERING
VALVE

THRUST
SLEEVE

FLYWEIGHT

FIG. 12-19. A centrifugal (flyweight) governor. (*Courtesy* Deere and Company)

usually not a problem. Several starting aids are available for cold-weather starting. Some tractors are equipped with electric glow plugs to heat the intake manifold before starting the engine. The heated intake manifold helps warm the air entering the engine so that the engine will start when cranked by the starter.

Ether may be used as a starting aid. Ether has a low ignition point so that the heat of compression in cold weather is sufficient to ignite the ether. This will then ignite the fuel charge. When the engine starts ether is no longer necessary.

Ether is available in spray cans designed for engine starting. Ether is highly explosive and should be sprayed into the engine only while the starter is cranking the engine. Too much ether can do serious damage to the engine!

Crankcase-oil heaters and engine-block heaters can also be used as starting aids. These heaters are usually elec-

trically operated and heat either the oil in the crankcase or the engine coolant.

SUMMARY

Diesel engines may be either of the two-stroke cycle type or the four-stroke cycle type. Both are used on farm tractors.

In diesel engines the fuel charge is ignited by the heat of compression. The fuel charge is injected by an intricate, high-pressure fuel-injection system. The fuel is kept clean by a dual filtering system.

The combustion chamber of the diesel engine must be designed to give turbulence to the fuel as it is injected in order to insure proper combustion of the fuel charge. Direct-injection, precombustion-chamber and auxiliary-chamber designs are used.

The governor on a diesel engine regulates the amount of fuel that is injected for each fuel charge by limiting the length of the fuel-injection stroke of the pump.

Most farm diesels use an electrical starting system similar to that used in spark-ignition engines. However, some diesel engines use an auxiliary gasoline-burning starting engine to crank the diesel engine.

Shop Projects

A. Removing water and sediment from the fuel system

Some diesel-fuel systems are equipped with a water trap, a sediment bowl, or both. These units are usually located between the fuel tank and the injection pump. Water is slightly heavier than diesel fuel and it will settle to the bottom.

1. Servicing the water trap. (See the operator's manual for details.)

    a. Open the drain cock at the bottom of the water trap and let the water drain out. (This should be done in the morning before starting the engine or after the tractor has not been in use for several hours.)

    b. Close the drain cock when all the water has been removed.

    c. Some fuel filters have drain cocks at the bottom. The water can be removed from these in the same manner as outlined above.

2. Servicing the sediment bowl. (See the operator's manual for details.) The general procedure is as follows:

    a. Close the fuel supply valve at the bottom of the fuel tank.

    b. Loosen the nut that holds the sediment bowl in place and move the holding bracket to one side.

    c. Remove the sediment bowl.

    d. Locate the gasket. It may be on the top of the bowl or with the bowl assembly. Do not damage the gasket. If necessary, use a new gasket.

e. Wash the bowl with kerosene or a safe solvent. Remove all sediment.

f. Install the bowl and gasket loosely.

g. Open the fuel-supply valve at the bottom of the tank and allow the sediment bowl to fill up with fuel. This forces air out of the bowl.

h. When the bowl is full of fuel, and all air is removed, tighten the holding nut so that the bowl seats firmly on the gasket.

B. Bleeding air from a diesel-fuel system

Whenever a diesel-fuel system is opened or has run dry, it is possible that air has entered the system. It is then necessary to bleed the system to remove the air. Air in the fuel system will cause hard starting, misfiring, and uneven power when the engine has started.

Diesel-engine fuel systems differ from one make of tractor to another. It is best to follow the instructions in the operator's manual for any specific engine. The flow of fuel is from the supply tank to filters, and then to the injectors. Air must be bled off at convenient points between the tank and injectors. In most systems, the first point where air can be removed is at the first filter. The next point is at the second filter, and the third point is at the injectors.

The general procedure is as follows:

a. Have the tank full of fuel. Open the fuel-supply valve at the bottom of the fuel tank so that the fuel can flow in the system.

b. Remove the bleed plug at the top of the first filter.

c. Allow the fuel to flow or pump the primer lever of the fuel pump until fuel with no air bubbles in it flows at this point. Replace the bleed plug.

d. Repeat the above for the second filter.

e. Start the engine.

f. If the engine still does not function properly, loosen the high-pressure fuel pipe at two or three injectors. (See the operator's manual for details on loosening the high-pressure lines for your engine.) Turn the engine over with the starter until fuel that is free of air bubbles flows from the loose connections.

g. Tighten the connections.

h. Repeat this for the remaining high-pressure lines. The engine should now run smoothly.

i. Check for leaks in the fuel system. Make necessary repairs.

Questions

1. What are the main differences between the diesel engine and the spark-ignition engine?

2. Why is the two-stroke cycle diesel engine more efficient than the two-stroke cycle gasoline engine? Give several reasons.

3. What is a precombustion chamber? Why is it used on some diesel engines?

4. Diesel fuel systems usually have two filters. Why is it so important to keep diesel fuel clean?

5. How do diesel starting systems differ from spark-ignition engine-starting systems?

6. How does the governing system on a diesel engine differ from that on a gasoline engine?

7. What are the commonly used starting systems for diesel engines?

8. What precautions must be observed when using ether as a cold weather starting aid?

9. Why are the parts in a diesel engine heavier than those in the spark-ignition engine?

References

Barger, E. L. *et al., Tractors and Their Power Units,* John Wiley and Sons, Inc., New York, New York, 1963.

Fundamentals of Service, Engines, Deere & Company, Moline, Illinois, 1977.

Fundamentals of Machine Operation, Tractors, Deere & Company, Moline, Illinois, 1974.

Jones, Fred R., *Farm Gas Engines and Tractors,* McGraw-Hill Book Company, New York, New York, 1963.

Tractor Maintenance, American Association for Vocational Instructional Materials, Athens, Georgia, 1975.

13   Cooling the Engine

A great deal of heat is generated in the combustion chamber of the engine as the fuel is being burned. Some of this heat goes into useful power, some is lost through the exhaust gases, and some must be removed by the cooling system of the engine. The cooling system handles about one-third of the heat generated. This heat passes through the cylinder walls into the cooling liquid or directly into the air, depending on the type of cooling system. The engine operating temperature must be carefully controlled. Permitting the engine to run too cold can be just as detrimental as running it too hot.

Most farm tractor engines operate with coolant temperatures in the 170° F to 200° F range. This is usually true for diesel engines as well as gasoline burning engines. Generally engines come equipped with the proper thermostat (Figure 13-2) to permit the engine to operate at the temperature recommended by the manufacturer.

LIQUID COOLING SYSTEMS

An engine equipped with a liquid cooling system has a water jacket around the cylinder walls and the cylinder head of the engine (Figure 13-1). The water jacket is connected to the radiator by passages and hoses. The engine is cooled by coolant circulating continuously from the water jacket through the radiator from top to bottom and back through the water jacket from bottom to top. The rush of air past the radiator cells or tubes picks up heat from the coolant. The fan keeps a constant flow of air past the radiator and helps in the cooling process.

Two similar types of liquid-cooling systems, the thermosiphon and the forced-circulation systems, are used on farm tractors.

The Thermosiphon System

The *thermosiphon system* is used in some engines, although it is not as common as it was during the earlier stages of the farm tractor. It is a simple system in which the circulation of coolant is caused by a difference in temperature between the coolant in the water jacket immediately around the combustion space and the coolant in the other parts of the cooling system. The coolant in the water jacket around the combustion space becomes hot, and, as it is being heated, it becomes lighter and rises to the top of the water jacket around the cylinder head and eventually to the radiator (Figure 13-1). In the radiator this hot coolant becomes cool and heavier so that its natural circulation is to the bottom of the radiator. It then enters the

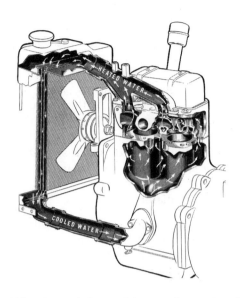

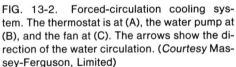

FIG. 13-1. A thermosiphon cooling system. It does not have a pump or a thermostat. (*Courtesy* Deere and Company)

lower part of the water jacket around the cylinders and is heated again by the hot area around the combustion space. The circulation is continuous as long as the engine is running. When the engine gets hot, the circulation is more rapid. As the engine cools, the circulation becomes slower. This system is quite efficient but it requires a larger radiator than does the forced-circulation system. Also, temperature control is not accurate.

The Forced-Circulation System

The *forced-circulation system* has all the parts found in the thermosiphon system, but in addition it has a water pump and a thermostat. The water pump is used to cause continuous circulation of the coolant. The thermostat controls flow of coolant through the cooling system and thereby regulates the operating temperature. The thermostat is an important unit in this system. It is located at the point where the coolant leaves the cylinder head (Figure 13-2).

FIG. 13-2. Forced-circulation cooling system. The thermostat is at (A), the water pump at (B), and the fan at (C). The arrows show the direction of the water circulation. (*Courtesy* Massey-Ferguson, Limited)

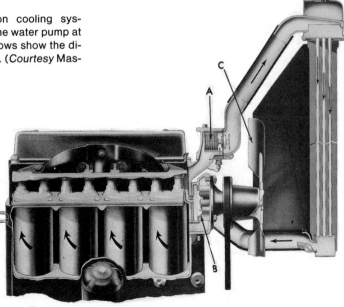

When the coolant reaches the proper temperature, the thermostat opens and allows it to circulate through the entire system. When the engine is cold, the coolant is bypassed and allowed to circulate only through the engine block. This causes a rapid warm-up of the engine.

Most forced-circulation systems are of the pressure type. They have a radiator cap with a relief valve which permits operating the cooling system under pressures of from about six to nine or more pounds per square inch (Figures 13-3 and 13-4). Each pound of pressure increases the boiling point of water about 3° F. A pressure-cooling system permits operating the engine at a higher temperature and reduces the loss of the coolant by evaporation. The pressure cap also has an atmospheric valve which permits air to go back into the cooling system as the engine cools. If this valve were not present, the radiator might collapse as the engine cools, since the pressure in the system is then reduced because of contraction. Pressure-cooling systems should not be filled completely. One or two inches of space should be left for expansion of the coolant.

The Radiator

The radiator is composed of cells, or tubes, through which the coolant is circulated. The fan immediately behind the radiator pulls the air through it and removes heat from the coolant. Figure 13-5 shows the cellular type of radiator. The tube type is shown in Figure 13-6.

The Water Pump

Forced-circulation systems use a water pump similar to that shown in Figure 13-7. The purpose of the pump is to speed up the circulation of coolant in the system and increase the rate of heat removal. The pump is usually driven by the fan belt or a similar belt provided especially for the pump. The water-pump bearing may be factory lubricated and not require any additional lubrica-

FIG. 13-3. Pressure-type radiator cap. (*Courtesy* Allis-Chalmers Manufacturing Company)

FIG. 13-4. Sketches showing the operation of a pressure-type radiator cap. (*Courtesy* Deere and Company)

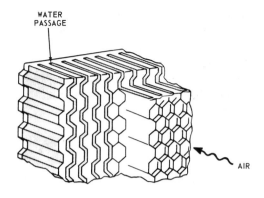

FIG. 13-5. Radiator of the cellular type. (*Courtesy* Deere and Company)

FIG. 13-7. A water pump for a tractor cooling system. The pump impeller is shown in white. (Photo by Authors)

tion or it may be the type that requires regular lubrication. The operator's manual should be consulted to determine the lubrication requirements for any particular pump. Figure 13-8 shows the parts of a water pump.

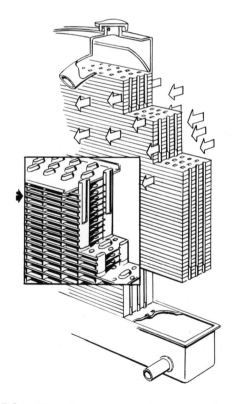

FIG. 13-6. Tube-type radiator construction. The arrows show the direction of the airflow. (*Courtesy* E. I. DuPont de Nemours and Company, Inc.)

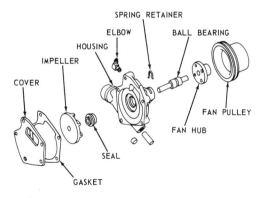

FIG. 13-8. An exploded view showing the parts of a water pump. (*Courtesy* Deere and Company)

The Thermostat

The thermostat used in most cooling systems is of the bellows type shown in Figure 13-9. The bellows are filled with a low-boiling-point liquid. When this liquid boils, it expands and causes the bellows to expand and open the valve. As the liquid in the bellows cools again, the bellows contract and close the valve. If the bellows become cracked or corroded, the liquid on the inside is lost, the valve stays open, and the thermostat is then of no value to the cooling system. It is a good idea to check the thermostat at

FIG. 13-9.   The thermostat at the top has a damaged bellows and is stuck open. The thermostat at the bottom is functioning properly. (Photo by Authors)

least once a year to make sure that it functions properly. A faulty thermostat should be replaced with a new one. If the tractor engine warms up slowly, it may be because the thermostat is stuck open. The temperature at which the thermostat will operate is usually stamped on the unit.

Cooling Fans

The fan is an important part of the cooling system. It increases the airflow through the radiator, so that the heat is removed more rapidly than it would be if there were no fan. Fans are usually driven by a fan belt or by gears. Belt-driven fans should be carefully adjusted for belt tension. A high belt tension will result in rapid bearing wear. If the belt is too loose, inadequate cooling may result and the belt may slip, causing rapid wear. When the generator and fan are driven by the same belt, the generator mounting bracket is usually adjustable to make it possible to get the proper belt tension. On some tractors, adjustable pulleys are used to secure proper belt tension.

AIR-COOLING SYSTEMS

Small stationary engines, garden tractors, and some power units are air-cooled. This system forces air past the hottest part of the engine block and cylinder head to pick up the excess heat and keep the engine at the proper operating temperature. The block and cylinder head are provided with fins to increase the surface from which heat can be passed on to the air. Sheet-metal shrouds are also provided to control the flow of air. The air is forced through the system by a fan that is usually a part of the flywheel. Figure 13-10 illustrates a typical air-cooled engine. It is very important to keep the screened inlet to the fan free of leaves and chaff that might cut down the flow of air. A reduced airflow will cause the engine to run too hot, which may result in permanent damage.

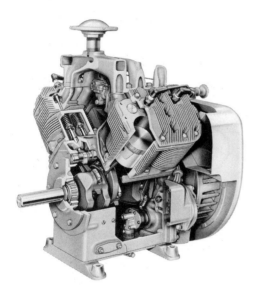

FIG. 13-10. Air-cooled engine. The fan at the right blows air over the engine. The fins around the cylinders aid in cooling. (*Courtesy* Wisconsin Motor Corporation)

ENGINE OPERATING TEMPERATURE

The cooling system of the engine is provided for the purpose of keeping the engine at its proper operating temperature. The entire system must be in good working condition in order to do its job effectively. If the engine is allowed to run too cold, rapid wear will take place. In the combustion process, approximately one gallon of water is formed for every gallon of fuel that is burned. If the

engine-operating temperature is too low, some of this water will condense in the cylinder and get past the piston into the crankcase and do harm to the engine. The cylinder walls will not be properly lubricated, and rapid engine wear will result.

Operating a tractor at light load during cold weather will cause the cooling water temperature to be too low unless special precautions are taken to keep the tractor operating at its proper temperature. A cover over the radiator, a good thermostat, and radiator shutters all help to secure the proper operating temperature. Figure 13-11 shows the effect of improper cooling water temperature on engine wear. The engine also has more power when it is operating at proper temperature.

Running an engine too hot is also detrimental, as it results in excessive wear and may cause warped and burned valves. An engine may be running hot for one of several reasons. Some of the more common ones are:

1. An accumulation of chaff or leaves on the radiator
2. An overload on the engine
3. Operation with insufficient coolant in the cooling system
4. Excessive lime deposits in the cooling system
5. Improper carburetor adjustment (usually too lean)
6. Incorrect ignition timing
7. Loose fan belt
8. Use of a fuel too low in antiknock quality for the engine

Regardless of the cause of overheating in the engine, it should be remedied before serious damage is done. Corrective measures for most of the causes of

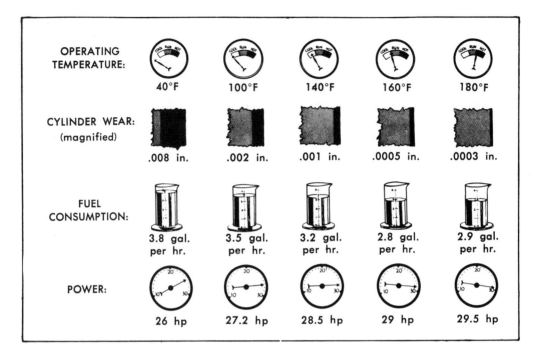

OPERATING TEMPERATURE:	40°F	100°F	140°F	160°F	180°F
CYLINDER WEAR: (magnified)	.008 in.	.002 in.	.001 in.	.0005 in.	.0003 in.
FUEL CONSUMPTION:	3.8 gal. per hr.	3.5 gal. per hr.	3.2 gal. per hr.	2.8 gal. per hr.	2.9 gal. per hr.
POWER:	26 hp	27.2 hp	28.5 hp	29 hp	29.5 hp

FIG. 13-11.   The results of tests show that as the operating temperature increases up to 180°F the cylinder wear and fuel consumption decrease and the power increases. (*Courtesy* American Oil Company)

overheating are quite obvious. However, excessive lime deposits are not easily removed. The prevention of lime deposits is easier than their removal. Use rainwater in the cooling system if there is any doubt about the well water that is available. Some commercial preparations are available that will aid in removing excessive lime deposits.

Loose scale in the cooling system can be flushed out with a garden hose and water under pressure. Remove all radiator hoses and the thermostat before flushing by this method. Force the water through the block, the cylinder head, and the radiator in a direction opposite to that of the normal flow of the coolant. Then replace the thermostat and the radiator hoses in the cooling system.

FIG. 13-12.   Use a brush to clean the front of the radiator. (*Courtesy* Massey-Ferguson, Limited)

ANTIFREEZE SOLUTIONS

Permanent type antifreeze (ethelene glycol) should be used in farm engine cooling systems when there is danger of freezing due to cold weather. Most engine manufacturers recommend that the antifreeze be left in the cooling system at all times—summer and winter. The antifreeze contains a corrosion inhibitor that helps to protect the cooling system from rust.

The cooling capacity or specific heat of antifreeze is not quite as high as that of water. For this reason it is sometimes recommended that water only be used during heavy load operation in the summer time. If a heating problem exists, removing the antifreeze and replacing it with water may help. However, other factors that may cause the engine to run too hot should also be checked. Follow the manufacturer's recommendations for servicing and maintaining the cooling system.

SUMMARY

Engines are cooled by liquid and air systems. The thermosiphon and forced-circulation systems are commonly used liquid cooling systems. The forced-circulation system is more common on farm tractors because it is efficient and readily controlled.

Air-cooling systems are found on small engines and on some power units. Air forced over the hot surfaces of the engine does the cooling.

Proper maintenance is highly important with all cooling systems. Keep the radiator clean on the inside as well as on the outside. Use soft water, the proper thermostat, and antifreeze if necessary. Keep the air intake clean on air-cooled engines. A properly maintained cooling system will insure proper operating temperature, good engine performance, and better fuel economy, and prevent excessive wear on the engine. The operator's manual should always be consulted for servicing the cooling system of a specific engine.

Shop Projects

A.   Testing a thermostat for accuracy

If there is any doubt about the accuracy of a thermostat, it can be tested. If found faulty, it should be replaced by a new one.

1.   Remove the thermostat from the engine. (See Figure 13-2 for the thermostat location and refer to the operator's manual.)

2.   If a thermostat valve is stuck open, a new unit should be installed.

3.   If a thermostat valve is closed when at room temperature, proceed as follows:

(a)   Determine the temperature at which the thermostat is to open. (This is usually stamped on the unit.)

(b)   Have available a thermometer that will register to about 200° F.

(c)   Place the thermostat and the thermometer in a pan of water. Make sure that the water covers the thermostat and keep the thermometer and thermostat off the bottom of the pan. (Figure 13-13)

(d)   Apply heat and notice the temperature at which the thermostat valve opens. This should be within a few degrees of the temperature that is stamped on the body of the thermostat. For example, a 165° F thermostat should open when the water reaches that temperature, but may not be fully open until the water has been heated to several degrees (10° to 15° F) above that temperature.

(e)   If the thermostat is satisfactory, it can be used again. If unsatisfactory, it should be replaced with a new one. When installing a thermostat in the engine, use a new gasket to avoid leaks.

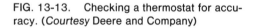
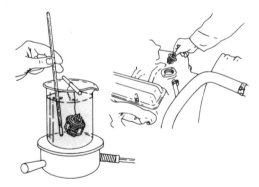

FIG. 13-13.   Checking a thermostat for accuracy. (*Courtesy* Deere and Company)

B.   Flushing a cooling system

Loose scale and other loose deposits can be flushed from a cooling system as follows:

1.   Open the drain plugs in the radiator and the engine block.

2. Remove the upper radiator hose and the thermostat. (See the operator's manual for instructions on removing the thermostat.)

3. Remove the lower radiator hose.

4. Force water through the block and the cylinder head by inserting a garden hose at the upper radiator-hose connection. Do this until the water coming out at the bottom is clean.

5. Force the water into the lower hose connection on the radiator until all the loose material has been flushed out. *Note:* If compressed air is available, it should be introduced into the water stream from the garden hose to make the flushing action more aggressive.

6. Test the thermostat, if necessary.

7. Inspect the radiator hoses and replace them with new ones, if necessary.

8. Install the thermostat and the hoses, close the drain plugs, and fill the cooling system with clean, soft water.

C. Special shop project on "trouble shooting."

In the previous chapters we discussed many of the basic principles and procedures involved in the operation and care of internal combustion engines. This is a good time to do a "trouble shooting" exercise. A suggested exercise is found in Appendix C.

Questions

1. How does a liquid cooling system help keep an engine at its proper operating temperature?

2. What is the proper coolant temperature for a gasoline-burning engine? A diesel engine?

3. Where should the thermostat be placed in a forced-circulation cooling system? Why?

4. What causes the coolant to circulate in a thermosiphon system?

5. What is a pressure cooling system? Why are pressure cooling systems used on some engines?

6. What are the advantages of an air-cooling system for an engine? What are the disadvantages?

7. How does engine operating temperature affect engine wear? Power?

8. How can a thermostat be checked for accuracy?

9. What are the most common causes of engine overheating?

References

Fundamentals of Service, Engines, 3rd Edition, Deere & Company, Moline, Illinois, 1977.

Fundamentals of Machine Operation, Preventive Maintenance, Deere & Company, Moline, Illinois, 1973.

Fundamentals of Machine Operation, Tractors, Deere & Company, Moline, Illinois, 1974.

Tractor Maintenance, American Association for Vocational Instructional Materials, Athens, Georgia, 1975.

14 Lubricating Oils and Greases

We commonly think of lubricating oil in the engine as a substance that minimizes wear and reduces frictional losses between moving parts. This chapter discusses the function of lubricants, as well as the properties of oils and greases. It is important to understand the six functions of lubricating oil in the engine, the type of friction found in internal-combustion engines, the classification of oils, oil additives, and the "when" and "why" of changing oil periodically.

TYPES OF FRICTION

Friction is the resistance to movement between two objects in contact with each other. If we had one steel bar sliding over another we would have *dry friction,* the direct contact of metal to metal. It is the purpose of the lubricant to separate the wearing parts.

If moving parts of the engine are separated by a thin film of oil or grease, allowing only the high points of the metal to touch each other, we have marginal lubrication. The oil or grease fills the openings in the metal, forming a cushion and holding the surfaces apart. This type of friction occurs in the engine after it has been stopped for a period of time and the oil has drained out of the lines. When the engine is started, more friction is present between the metal surfaces until oil pressure is delivered to provide viscous lubrication.

Viscous friction refers to the resistance to motion between adjacent layers of liquid. In the engine, oil will adhere to the metal surfaces of the moving parts. Since oil is delivered under pressure, engine bearings and other surfaces will float on a layer of oil eliminating any metal-to-metal contact and reducing the energy required to operate the moving parts. Viscous friction is the friction that occurs between the molecules of the lubricant, and it depends greatly on the viscosity or grade of oil being used. The lighter the oil, the less the viscous friction. To keep viscous friction and heat as low as possible, the recommended lighter oils should be used.

FUNCTIONS OF OIL

Most of us think that the only purpose of an oil in the engine is to reduce wear. However, oil has six important functions:

1. To minimize wear. It provides an oil layer in which only viscous friction occurs. It is obvious that if friction were present for any length of time,

rapid wear would cause failure of the engine.

2. To remove heat from engine parts. Since oil is circulated through the engine, it will absorb heat from the moving metal surfaces and carry this heat into the oil pan, where the cooler metal surfaces will absorb the heat from the oil and transfer this heat to the outside air. Thus, oil acts as a coolant.

3. To cushion metal parts in contact with each other. At certain times in the engine cycle, great amounts of pressure occur, tending to force metal-to-metal contact at the bearing surfaces. The oil film prevents this contact by absorbing the force that occurs on the power stroke.

4. To act as a cleansing agent. Since oil circulates, small metal particles, dirt, fuel carbon, and other materials are picked up or washed out of bearing surfaces and other parts of the engine. Most of the larger foreign particles will be carried to the oil filter and removed, or they may be deposited in the oil pan, where they will do no harm.

5. To act as a sealant. Oil is thrown off the crankshaft onto the cylinder walls, where it is controlled by the oil rings of the piston. This liquid fills in the irregularities in the metal of the cylinder walls and aids all the piston rings in providing a sealed chamber in which combustion can occur.

6. To reduce frictional losses. Approximately 5 percent of an engine's power is lost to internal friction. Oil present in sufficient quantities to provide viscous friction greatly reduces power losses due to friction between metal surfaces.

PROPERTIES OF OIL

Oil suitable for engine use must possess properties that will enable it to meet the many demands imposed on it by a working engine. Five important properties of oil are viscosity, viscosity index, pour point, color, and additives. We shall discuss each of these.

Viscosity

Viscosity is the tendency of an oil to resist flowing. An oil must have sufficient viscosity to resist being squeezed out from between metal parts under heavy load. Oil must be sufficiently fluid to flow easily through the oil lines and deliver the cushioning and coolant services demanded of it. To help us select oils of proper viscosity, SAE Viscosity Classification numbers are used. The viscosity of an oil is measured in Saybolt Universal Seconds, that is, the number of seconds required for a given amount of oil to flow through an opening at a given temperature. Saybolt Universal Viscosity is usually determined at temperatures of zero and 210° F. Summer-grade oils, such as SAE 20 and SAE 30, are based on viscosities at 210° F; winter-grade oils, such as SAE 5W and 10W, are based on viscosities at both zero and 210° F.

Viscosity Index

The extent to which viscosity will vary with temperature changes depends on the composition of the base oil as well as the additives used in the finished product. How much or how little viscosity varies with temperature change is indicated by the oil's viscosity index.

A viscosity index number (V.I. number) is helpful in selecting an oil for a

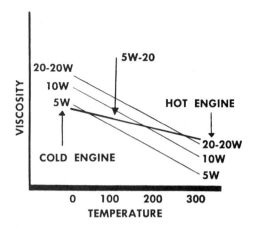

FIG. 14-1. Viscosity-temperature characteristics of a 5W-20 multi-grade motor oil compared with ordinary motor oils. The 5W-20 oil has a high viscosity index. (*Courtesy* American Oil Company)

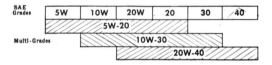

FIG. 14-2. Multigrade motor oils span several SAE (single) grades. (*Courtesy* American Oil Company)

specific use or engine. An oil that has a higher viscosity index number, such as 140, will have a more stable viscosity with temperature changes than one that has a lower viscosity index number such as 90. Generally speaking all oils tend to become more fluid, or less viscous, when hot and less fluid, or more viscous, when cold. Multigrade oils, such as SAE 10W-40, have a high viscosity index number. Single grade oils, such as SAE 30, will generally have a lower V.I. number.

Pour Point

This is a measure of an oil's ability to flow at low temperatures. One method of determining pour point is to take a small sample of oil, place it in a container and subject it to low temperatures. As the oil nears the point where it will not flow readily, a reading is taken at each 5° drop in temperature. When the container is tipped slightly and the oil does not move or tend to flow, the temperature taken immediately before this point is called the pour point of this oil and is expressed in ° F.

In cold climates an oil with a low pour point (−40° F) is desirable to facilitate starting the engine and to help provide oil circulation or lubrication until the engine approaches the proper operating temperature.

Color

The quality of an oil cannot be judged by its color. There are several conditions that cause an oil to change color after being in the engine for a short period of time. Water produced in the engine or taken in with the air through crankcase ventilation will cause the oil to take on a milky color. Finely divided lead ash from the antiknock material, tetraethyl lead, found in gasolines and deposited in combustion chambers will often find its way into the oil. This will give the oil a gray color. Carbon picked up from engine parts will impart a black color to oil. If the manufacturer's suggested oil-change period is followed, it is doubtful that these discolorations will cause harm to the engine. Crankcase oil should be changed more often during cold weather or during operation in dusty areas than in warm weather and in areas having less dust in the air.

Additives

Certain chemical additives are carefully blended with modern engine oils to prevent the formation of engine deposits, retard wear of certain engine parts, and aid in cleansing the engine. Oxidation inhibitors reduce the breakdown of oil at high temperatures and thus retard the formation of varnish and acids. Corrosion inhibitors prevent the formation of acids. Antiwear agents retard the wear of piston rings and components of the valve assembly. Detergent-dispersants clean engine surfaces and keep the deposit-forming materials in suspension. Antifoam agents prevent the crankshaft churning action from causing the oil to foam up and be lost through the crankcase ventilating system. Additives help keep the oil in good condition to do its normal lubricating job.

OIL CLASSIFICATION

Viscosity classification has been discussed as a method of selecting the proper oil. Another method of classification, developed by the American Petroleum Institute (API), enables the operator of an engine to judge the quality of oil.

In 1971 the API put into effect a new oil classification system. This system is more flexible than the previous system that was used. A heavy duty oil may meet or exceed the requirements of several of the API classifications. Also, it is now possible to add new classifications as demand arises without upsetting existing classifications. The 1971 API oil service classifications and descriptions of service that they are intended to give are as follows:

1. SA—Oil in this classification does not offer the engine protection afforded by compounded oils. Oil additives designed to protect engines are not included.

2. SB—These oils are designed for engine operation under mild conditions and provide only minimum protection against engine wear.

3. SC—This class of oil provides control of high and low temperature deposits, wear factors, and rust and corrosion protection. It is intended for use in passenger car engines and provides moderate engine protection.

4. SD—Designed primarily for use in passenger car engines, this oil provides more protection than the SC oils.

5. SE—This oil is intended primarily for use in passenger cars and provides high-temperature anti-oxidation protection, low-temperature anti-sludge, and anti-rust protection. Oil in this classification provides excellent protection for gasoline engines.

6. CA (Light Duty Diesel Engine Service) This is suitable for gasoline engines and diesel engines operated in mild to moderate duty with high quality fuels.

7. CB (Moderate Duty Diesel Engine Service) This oil is used in gasoline engines and moderate duty diesel engines operating on lower quality fuels which require more protection from wear and deposits.

8. CC (Moderate Diesel and Gasoline Engine Service) This class is suitable for diesel engines operated on severe duty and includes heavy-duty service gasoline engines.

9. CD (Severe Duty Diesel Engine Service) This oil is recommended for supercharged diesel engines in high-speed, high-output duty requiring

highly effective control of wear and deposits.

An oil container should show the API classification of its contents. Some will show that the oil is suitable for more than one type of service, for example SE and CB. This means that the oil will provide excellent lubrication for spark ignition engines (SE) and for diesel engines operating in relatively good conditions (CB). Heavy duty (HD) oils generally meet or exceed several API service classifications, and are suitable for use in diesel as well as spark ignition engines. However, the operator's manual should be consulted for the proper oil classification for a specific engine.

WHEN TO CHANGE OIL

You may have heard that oil does not wear out. This is true in the sense that oil does not lose its lubricating qualities. However, many contaminants enter the oil soon after it is placed in service. Four harmful contaminants are: water, sludge, varnish, and acids. How quickly the oil becomes contaminated with these products depends on engine-operating conditions. It might appear that all metal particles that wear off the engine under normal service conditions are filtered out of the oil. This is not true, although most of them will be taken out. If the oil is allowed to remain in service longer than recommended, large amounts of very fine metal particles will build up and be carried by the oil throughout the lubricating system, causing excessive wear, or will be deposited where they will have harmful effects. How often oil should be changed will vary with climatic conditions, service

conditions, the mechanical condition of the engine, and other factors. Following the manufacturer's recommendations and knowing the conditions under which the engine must operate will be your best guides for oil change. Remember—oil must be changed regularly to reduce wear and to prolong the life of an engine. Special additives used in high-quality oils gradually become depleted, reducing the protection to engine parts.

TYPES OF GREASE

Grease is basically a lubricating oil with a soap-type thickening agent added to give it a thick consistency. Lime soap is used in chassis grease and results in a water-resistant product that generally can be used anywhere that high temperatures are not present. Soda soap is used to form a semismooth product often referred to as wheel-bearing grease. This lacks the water resistance of lime-soap grease but possesses heat resistance. Lithium soap combines the water resistance and heat resistance of the other greases to give a multipurpose lubricant that is suitable for all-round use on farm machinery. This grease makes it possible to handle all fittings with one grease gun.

Grease is used to lubricate bearings when they are not generally accessible or not tightly sealed, or when a single lubrication must last for a rather long period of time. Grease also serves to exclude dirt and water.

Grease fittings should be wiped with a clean rag before being serviced, to prevent dirt from being forced into the bearing. Enough grease should be used in bearings that are not tightly sealed to force some of the surface material out and off the bearings, thus cleansing them.

Wheel bearings should be serviced as recommended by the manufacturer. Frequent servicing is usually advised when equipment is being operated in extremely dusty or wet surroundings. In most cases, the bearings should be removed from the spindle. Thoroughly clean the bearings in kerosene, or other suitable solvent, removing all the old grease. Examine them for signs of chipping or excessive wear. Hand pack a cleaned bearing by placing a quantity of grease in the palm of your hand and forcing the grease into the open end of the bearing by squeezing the grease between the bearing and the heel of your hand. It is important to clean all the old grease from the wheel hub and spindle and to prevent dirt or other contamination from getting on the newly greased bearing, spindle, or hub on reassembly. The importance of cleanliness and freedom from contamination cannot be overstressed.

SUMMARY

Friction between moving engine parts results in power losses and wear to components of the engine. Oil and grease are used to overcome this friction and reduce wear, thus causing long life and trouble-free performance in engines. Oil has functions other than reducing friction and retarding wear: (1) removing heat from engine parts, (2) acting as a cushioning agent between metal parts, (3) cleaning parts and carrying impurities to the oil filter, and (4) acting as a sealant.

Selecting the proper oil for an engine is very important and should not be overlooked. Always use clean oil that has been properly stored. Five important properties of oil are: (1) viscosity, (2) viscosity index, (3) pour point, (4) color, and (5) additives.

Viscosity classification of oil enables us to select the proper oil for the engine and for the conditions under which it will operate. Follow the manufacturer's recommendations for the proper SAE grade to use in the various climates where there are any temperature variations.

API service classifications of oil give us a method of selecting oil to provide proper protection for the engine parts under varying load conditions. All oil containers should be marked with the viscosity and service classification. The 1971 API oil classifications are SA, SB, SC, SD, SE, CA, CB, CC, and CD.

Many theories are advanced by engine operators to determine when engine oil should be changed. Because oil becomes contaminated, periodic oil changes are needed. Follow the manufacturer's recommendations on oil changes. However, it may be necessary to adjust this period according to the climatic conditions, the service conditions, and the mechanical condition of the engine.

Additives are used to add desirable properties to oil. Actually, grease is an oil, thickened by a soap-type agent so that it will stay where it is placed. It tends to exclude air from bearing surfaces and prevent oxidation. Water- and heat-resistant properties of greases are important considerations when selecting this lubricant. Generally, soda-soap, lime-soap, and lithium-soap greases are used. Multipurpose (lithium soap base) greases are by far the most common for farm machinery.

Shop Projects

A. Lubrication principles

1. Obtain two flat pieces of metal.

2. Place them together, then try to move one piece while applying pressure.

3. Spread a small amount of oil or grease on one piece; then apply pressure to the two pieces while trying to move one. This will illustrate a fundamental principle of lubrication.

B. Oil classification

1. Obtain several empty oil containers (oilcans) from service stations.

2. Examine the oil containers for the API classification letters while observing the viscosity classification.

3. Note that some companies will list all the API classifications, while others will list only the highest classifications.

C. Packing front-wheel bearings

1. Jack up the front wheel of a tractor and insert a block as a safety precaution.

2. Remove the hub cap, the retaining nut, the outer bearing, and the wheel.

3. Remove the inner bearing, the seal, and the retainer. Be careful not to damage the inner cup or bearing.

4. Remove all old grease and wash the hub and bearings in a safe solvent. Dry, but do not spin, the bearing with compressed air. Spinning is dangerous and may damage the bearing.

5. Check the bearings and the cup for wear and replace them, if necessary.

6. Pack the bearings by placing grease in the palm of the hand and forcing it into the bearing. All spaces between rollers should be filled.

7. Check the seals and the retainer and replace them if they are worn.

8. Replace the inner cup, the bearings, the retainer, and the seals.

9. Replace the wheel.

10. Adjust the bearings by drawing up the nut until the wheel begins to drag; then loosen the nut 1/6 to 1/4 of a turn so that the wheel turns freely but has no end play. Replace the cotter pin.

11. Refer to the operator's manual when tightening the bearings.

Questions

1. List the six functions of oil in the engine. Why are these functions important?

2. List the five properties of oil. Explain each.

3. What is the 1971 API oil classification system?

4. When should oil be changed?

5. List the three common types of grease and the properties of each.

6. Which grease is most commonly used for farm machinery? Why?

References

Fundamentals of Service, Engines, 3rd Edition, Deere & Company, Moline, Illinois, 1977.

Fundamentals of Machine Operation, Tractors, Deere & Company, Moline, Illinois, 1974.

Tractor Maintenance, American Association for Vocational Instructional Materials, Athens, Georgia, 1975.

15   Lubricating Systems

After considering the subject of friction in Chapter 14, we should know that viscous friction must be provided wherever possible in the engine. If oil is to act as a lubricant, coolant, cushioning agent, sealant, and cleansing agent, an extensive lubricating system is necessary in order to distribute the oil to the moving parts of the engine. The discussion in this chapter will cover the lubricating systems, their principles of operation, and their component parts. Oil filtering and crankcase ventilation are important topics to consider when studying lubricating systems.

CIRCULATING-SPLASH SYSTEM

This system uses a combination pumping-and-splashing action to lubricate the moving parts of the engine. The churning action of the crankshaft and connecting rods splashes oil to all moving parts in the crankcase. The connecting-rod caps often contain oil scoops or dippers to gather and direct oil to the connecting-rod bearings. Splash-pan troughs are located directly below each connecting rod and are filled with oil from a pump located in the engine pan. Some of the older engines did not use a pump, but relied on gravity flow to keep the troughs supplied with oil.

A fine oil mist fills the crankcase from the churning action of the crankshaft. This mist lubricates the moving parts in the upper crankcase.

FULL-PRESSURE SYSTEM

The full-pressure oil system uses drilled passageways and oil lines to carry oil under pressure from the oil pump to bearing surfaces and other engine parts requiring lubrication. Oil escaping from these bearing surfaces is thrown off by the crankshaft against the cylinder walls, thus furnishing lubrication. Figure 15-1 illustrates the full-pressure system.

The engine lubricating oil is taken by the oil pump from the oil pan through a floating, screened intake. This screen takes oil near the surface of the oil in the pan; therefore, less sediment and dirt are picked up and circulated by the pump. The floating screen is pivoted and is free to move up and down with the oil level. Oil is delivered to an oil gallery, under pressure, that is controlled by a spring-loaded regulator valve in the pump body. Oil from the gallery is delivered through drilled passages to the camshaft bearings, the crankshaft main bearings, and the connecting-rod bearings. Other passages and oil lines carry oil to the valve mechanism, the engine governor, and other parts requiring lubrication.

OIL PUMPS

Oil pumps are of two types, the gear and the vane. The gear pump is most widely used.

The Gear Pump

The gear pump consists of two gears in mesh inside a close-fitting case. One gear of the oil pump is driven by a shaft connected to the engine camshaft. Oil is taken into the pump on the side of the gears that is going out of mesh. The oil is carried around the outside of the gears and discharged into the pump outlet.

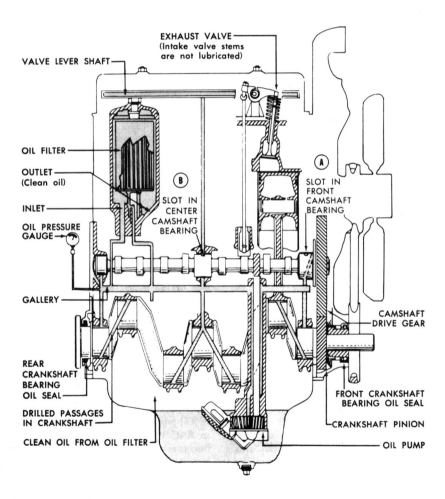

FIG. 15-1.   The oil-flow and parts of a tractor full-pressure lubricating system showing how oil is delivered to each part of the engine. (*Courtesy* International Harvester Company)

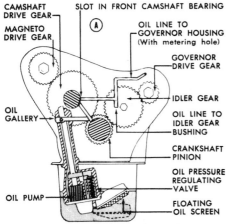

Section through front of engine showing how
oil is intermittently forced to idler gear
bushing and governor

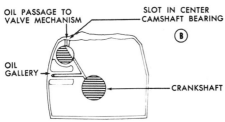

Section through center of engine showing how
oil is intermittently forced to valve mechanism

FIG. 15-1 (Continued)

The Vane Pump

The van-type pump uses movable vanes mounted within a rotor. The rotor is driven, causing the vanes to turn with it. Oil-pump vanes should be installed so that the flat sides will face in the direction of rotation. Oil enters the wide area between the rotor and the housing and is carried by the extended vane to a point where there is little clearance between the rotor and housing. This action forces the vane to slide into the slot in the rotor and at the same time forces the oil picked up by the vane into the oil outlet line. Figure 15-2 shows vane-type and gear-type oil pumps.

OIL FILTERS AND THEIR FUNCTION

Most engines are equipped with an oil filter. Its purpose is to trap and hold abrasive materials that would otherwise get into the bearings, causing rapid wear. Filters are of two types, partial-flow and full-flow. In the partial-flow system, only a small portion of the oil is filtered as it circulates through the engine. In the full-flow system, all of the oil is filtered as it circulates through the engine. This system incorporates a bypass valve that allows the oil to circulate past the filter when the engine is cold or if the filter should become clogged. The by-pass valve found in the full-flow system is shown above the filter in Figure 15-3. This spring-loaded valve can be forced open to permit oil circulation past the filter.

CRANKCASE VENTILATION

Ventilation systems are used on most engines to dispose of the unburned fuel vapor and moisture vapor that accumulate in the engine crankcase. This system prevents the formation of sludge and other contaminants, particularly during warm-up periods and during cold-weather operation. Figure 15-4 shows the engine breather cap of the ventilation system.

Outside air is taken in through the engine breather by the vacuum produced in the crankcase from the upward movement of the pistons toward TDC. As the pistons move toward BDC, a pressure is created that forces the mois-

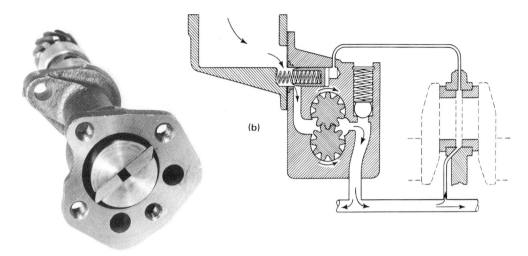

FIG. 15-2.   (A) Bottom view of the vane-type oil pump shows the vanes in the partially extended position. The oil is picked up by a vane as it begins to move outward in the rotor slot, is carried around the housing, and is forced out of the pump as the area occupied by the oil becomes restricted, and the vane moves inward in the rotor slot. (*Courtesy* White Motor Company) (B) In the gear-type pump, the arrows show the flow of oil through the gear mechanism and out through the oil lines to the moving engine parts. The spring-and-ball type of relief valve is shown. This relief valve enables the oil pump to continue operation without high back-pressure if the oil line becomes clogged or when the oil is cold and cannot be forced through the oil lines as rapidly as the pump is delivering oil. (*Courtesy* Caterpillar)

ture and vapor out of the crankcase through a crankcase vent pipe. This ventilating action prevents a pressure build-up in the crankcase, and eliminates harmful vapors and moisture. Regular servicing of the breather cap and the crankcase ventilation tube is necessary for proper engine maintenance.

A pressure-type ventilating system uses a vane-type pump. Air from the air cleaner is pumped into the crankcase, forcing harmful vapors out and into the intake manifold, where these vapors are taken into the engine and burned.

SUMMARY

Circulating-splash and full-pressure lubricating systems are used on modern engines. The full-pressure system is now in more general use.

The gear-type oil pump is used on most engines. Since it has few moving parts, this pump is a trouble-free, dependable unit. Partial-flow and full-flow oil filtering systems remove harmful material from the oil and help reduce engine wear. All filtering systems should contain a bypass valve to allow oil flow to continue should the filter become inoperative.

Crankcase ventilation is accomplished by taking air in through the engine breather cap. This contains a screen to remove coarse material from the air before it is taken into the engine. The screen on the breather cap should be cleaned periodically.

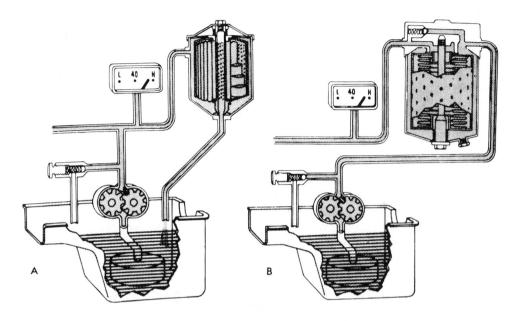

FIG. 15-3.   In the partial-flow system (A), the oil may go directly to the engine with-out being filtered. In the full-flow system (B), the oil must pass through the filter before going to the engine, except when the oil is cold or the filter becomes clogged. (*Courtesy* Standard Oil Company of Indiana)

FIG. 15-4.   The engine-breather cap of the ventilation system. (*Courtesy* Allis-Chalmers Manufacturing Company)

Shop Projects

A. Examining an oil pump

    1. Obtain a discarded oil pump from a service station or garage.

    2. Disassemble the pump and examine the simple construction of this unit.

    3. Reassemble the pump.

B. Examining oil-filtering systems

    1. Examine the oil-filtering system on a full-flow and a partial-flow installation.

    2. Check the cost of the filter for each system.

    3. Note that the full-flow system requires a high-quality filter to withstand oil-system pressure without breaking down.

C. Changing the engine oil and filter

    1. Be sure the engine is hot before removing the plug and draining the oil.

    2. Let the oil drain completely and replace the plug.

    3. Change the oil filter element. Be sure to wipe off the inside of the filter case before installing the new filter, the new gaskets, and the new seal rings.

    4. Replace the filter cap and tighten it firmly.

    5. Clean the dust and dirt from around the oil-fill pipe and fill the crankcase to the proper level with new oil of the type recommended for the engine.

    6. Start the engine and run it for a few minutes to check for oil leaks at the filter.

    7. Stop the engine and check the oil level.

Questions

1. How does the circulating-splash system differ from the full-pressure lubricating system?

2. What type of oil pump is most widely used? Why?

3. Describe the difference between partial-flow and full-flow oil-filtering systems.

4. What might be indicated by a very high oil pressure?

5. What might cause the oil pressure to be low?

6. Why is a crankcase ventilating system necessary?

7. What service is necessary to maintain the crankcase ventilating system?

References

Fundamentals of Service, Engines, 3rd Edition, Deere & Company, Moline, Illinois, 1977.

Fundamentals of Machine Operation, Tractors, Deere & Company, Moline, Illinois, 1974.

Tractor Maintenance, American Association for Vocational Instructional Materials, Athens, Georgia, 1975.

16 Tractor Types and Trends

Modern farm tractors have undergone continual changes and improvements to become the up-to-date and effective agricultural power units of today. New developments have helped to make these tractors more efficient, versatile, safe, convenient, and powerful.

Current tractors may be classified according to traction members used, and according to use and size of the tractors.

TRACTOR TYPES
(According to traction members)

Tractor types may be classed according to wheel or track systems. Classification may include:

1. Wheel Tractors

    (a.) Tricycle or three-wheel
    (b.) Four wheel (Two-wheel drive)
    (c.) Four wheel drive

2. Track type tractors

Wheel Type Tractors

The wheel type tractor is by far the most widely used tractor for agricultural purposes. Many wheel type tractors have been commonly made in the past as three-wheel or tricycle type tractors although the front or steering member may have two closely spaced wheels.

These tractors were designed as all-purpose tractors especially for row-crop use. Most tractors presently manufactured are of the four wheel type, with both front and rear wheel treads adjustable for use in row crops. Some are available with the narrow or row crop front as an option. The four wheel drive tractor, with propelling power delivered through both front and rear wheels, is gaining in popularity.

Track-Type Tractors

Track-type tractors are propelled by two heavy metal devices known as tracks. They are steered by controlling the speed of these tracks. Track-type tractors are not widely used in agriculture, but are well adapted to hilly areas and some tasks such as earth moving and land clearing.

TRACTOR TYPES
(According to use and size)

Tractors may also be classified according to use and/or size as follows:

1. Utility tractors
2. Large Field tractors

    (a.) Two wheel drive
    (b.) Four wheel drive

3. Orchard and Vineyard tractors

FIG. 16-1.   Small utility tractor. (*Courtesy* International Harvester)

4.   Lawn and Garden tractors
5.   Industrial tractors

Utility Tractors

The utility tractor is commonly manufactured in sizes up to about 80 h.p. These tractors may be the major source of power for field work on smaller or medium sized farms. They vary in size from the small general purpose or utility tractor, shown in Figure 16-1, to the 70 h.p. tractor, shown in Figure 16-2, and larger. Utility tractors may be used for a large variety of field and farmyard applications including tillage, harvesting, haying, loading, and other uses. Many of them (Figure 16-3) are

FIG. 16-2.   Utility tractor, 70 h.p., pulling baler. (*Courtesy* Deere and Company)

FIG. 16-3.   Utility tractor, 65 h.p.; can be used for row crops. (*Courtesy* International Harvester)

adapted to row crop use because of the adjustable front and rear wheel tread widths and higher clearance.

Large Field Tractors

Two Wheel Drive. Large two wheel drive (Figure 16-4) tractors are manufactured in the power range of approximately 85 to 180 h.p. These tractors are adapted to a wide range of heavy tillage and other field work, in addition to farmstead and other applications. Some of them have adjustable wheel tread widths and high clearance for row crop use. Some are available with very high clearance for such crops as sugar cane (Figure 16-5).

FIG. 16-4.   Large 2-wheel tractor doing tillage. (*Courtesy* Allis-Chalmers Manufacturing Company)

FIG. 16-5.   Tractor with extra-high clearance for sugar cane. (*Courtesy* Deere and Company)

Four Wheel Drive. The large four-wheel drive tractors (Figure 16-6) manufactured in sizes of up to 400 h.p., and in some cases larger, are gaining rapidly in popularity. These powerful tractors with excellent traction are well adapted to field work on large farms. Some two-wheel drive tractors are also available with optional front wheel power drive (Figure 16-7). Four wheel drive tractors may also be used for row crop work (Figure 16-8).

Orchard and Vineyard Tractors

Some tractors are designed especially for orchard (Figure 16-9) and vineyard work (Figure 16-10). These tractors are designed to have a low and smooth profile to avoid snagging branches and vines. The vineyard tractor is quite narrow.

FIG. 16-6.   Large 4-wheel drive tractor doing field work. (*Courtesy* Steiger Tractor, Inc.)

FIG. 16-7.   Two-wheel drive tractor with optional front wheel power drive. (*Courtesy* Deere and Company)

FIG. 16-8.   Four-wheel drive tractor cultivating row crops. (*Courtesy* J. I. Case)

FIG. 16-9.   Orchard tractor with low profile. (*Courtesy* Deere and Company)

FIG. 16-10.   Vineyard tractor, narrow with low profile. (*Courtesy* Deere and Company)

Lawn and Garden Tractors

Several manufacturers produce tractors designed for garden and lawn operations. The tractor shown in Figure 16-11 is designed for either lawn or garden use. The tractor shown in Figure 16-12 is a diesel garden tractor with power to the front wheels as well as to the rear wheels. Tractors are also manufactured for grounds maintenance (Figure 16-13), and with a very low center of gravity for highway mowing (Figure 16-14).

FIG. 16-11.   Lawn and garden tractor. (*Courtesy* FMC Corporation, Outdoor Power Equipment Division)

FIG. 16-12.   Diesel powered garden tractor. (*Courtesy* FMC Corporation, Outdoor Power Equipment Division)

FIG. 16-13.   Grounds maintenance tractor. (*Courtesy* Ford Motor Corporation)

FIG. 16-14.   Low center of gravity tractor for highway mowing. (*Courtesy* J. I. Case)

Industrial Tractors

Many types of heavy duty industrial tractors are produced for a variety of purposes. Some have agricultural uses such as the self-propelled loader shown in Figure 16-15. These vary in size from the tractor shown to very large sizes with heavy duty mounted loaders (Figure 16-16).

There are also several other types of specialty tractors manufactured such as those designed to provide power units for swathers. Another is designed so the seat and operator's console can face in either direction (Figure 16-17).

A large variety of tractor types and sizes are manufactured to meet the needs of modern agriculture. Most of the tractors are designed for a wide variety of uses, while some are developed for specialized applications.

Modern farm tractors are continually being improved to make them more efficient, safe, effective, and comfortable to operate. These developments are causing these tractors to become increasingly complex. Tractor types are designed and built to carry out the many and varied power requirements for agriculture. This results in several different types or variations of tractor types. Even with these substantial changes and improvements, much of basic tractor design has remained similar.

FIG. 16-15.   Small-size loader tractor. (*Courtesy* Dynamic Industries)

FIG. 16-16.   Larger industrial tractor with mounted loader. (*Courtesy* J. I. Case)

FIG. 16-17.   Tractor with reversible seat and console. (*Courtesy* Versatile)

TRACTOR TRENDS

Many interesting and important trends are developing in the design, manufacture, and use of tractors. A brief discussion of several of these trends is presented here.

Number of tractors

As shown in Table 16-1 there has

TABLE 16-1. PRODUCTION OF TRACTORS (in units)

Year	Wheel type & Tracklaying	Wheel type	Tracklaying	Garden Tractors† & Motor Tillers
1925–29 Avg. .	185,727	172,179	14,091	4,665
1936–39 Avg. .	222,161	197,445	24,716	9,095
1940–44 Avg. .	249,336	217,874	31,463	12,987
1945–49 Avg. .	442,648	404,230	38,419	126,121
1950	542,448	498,768	43,680	151,198
1951	617,060	567,446	49,614	177,169
1952	463,211	414,560	48,651	199,321
1953	442,247	390,385	51,862	245,609
1954*	285,159	245,755	39,404	191,235
1955	377,114	330,141	46,973	185,696
1956	272,265	214,654	57,611	200,997
1957	265,852	229,050	36,802	178,542
1958	265,495	241,269	24,226	227,073
1959	297,078	259,916	37,162	327,083
1960	178,500	152,187	26,313	416,455
1961	191,191	171,417	19,774	389,861
1962	207,281	188,101	19,180	421,833
1963	224,306	203,449	20,857	347,413
1964	240,579	213,221	27,358	416,486
1965	271,751	244,050	27,701	422,342
1966	299,114	270,687	28,427	507,606
1967	262,061	242,215	19,846	560,387
1968	236,023	213,199	22,824	583,265
1969	218,547	195,704	22,843	529,746
1970	191,679	171,603	20,076	498,372
1971	186,249	167,501	18,748	491,117
1972	218,213	196,988	21,225	662,583
1973	236,944	212,072	24,872	847,427
1974	233,825	210,074	23,751	1,287,026
1975	231,968	210,913	21,055	1,287,097
1976	213,154	193,724	19,430	1,109,978

* Figures subsequent to 1954 exclude contractors' off-highway wheel type.
† Figures represent manufacturers' shipments subsequent to 1957. Production not available on comparable basis.
Data: Current Industrial Reports: Series M35S and M35A, Bureau of the Census.
(**Courtesy** Implement and Tractor Intertec Publishing Corporation)

been a decrease in numbers of wheel-type and track-laying tractors produced since 1973. Wheel type tractor production declined from 212,072 in 1973 to 193,724 in 1976. The number of tractors on farms, excluding garden tractors, has also declined steadily from 4,778,000 in 1963 to 4,397,000 in 1977 (Table 16-2). This decline is offset by a substantial increase in the size of tractors being produced.

TABLE 16-2.   NUMBER OF WHEEL AND CRAWLER TRACTORS ON U.S. FARMS, DESIGNATED YEAR

1953	4,100,000
1954	4,243,000
1955	4,345,000
1956	4,480,000
1957	4,570,000
1958	4,620,000
1959	4,673,000
1960	4,688,000
1961	4,743,000
1962	4,763,000
1963	4,778,000
1964	4,786,000
1965	4,787,000
1966	4,783,000
1967	4,786,000
1968	4,766,000
1969	4,712,000
1970	4,619,000
1971	4,562,000
1972	4,509,000
1973	4,477,000
1974	4,478,000
1975	4,463,000
1976	4,434,000
1977	4,397,000

SOURCE: Reports of the USDA and the Bureau of the Census. After 1959 Alaska and Hawaii are included. On-farm data after 1970, except for 1974, are estimated by I&T.
(*Courtesy* Implement and Tractor Intertec Publishing Corporation)

Horsepower Trends

Tractor size and power has been increasing dramatically in recent years. An examination of Table 16-3 reveals that the production of tractors in the 35-39 horsepower category dropped from 23,974 in 1970 to 7,499 in 1976. The production of tractors rated at 140 h.p. or more increased from only 1,325 in 1970 to 6,191 in 1972. In 1974 there were 16,951 two-wheel drive tractors produced in this power category and in 1976 this number increased to 18,221, the sum of three different categories at or above 140 h.p.

Trend to Four-Wheel Drive Tractors

Large four-wheel drive tractors are now popular and are being produced in large numbers. As indicated in Table 16-3, there were 8,287 of these produced in 1974 with 1,473 up to 169 h.p. and 6,814 of 170 h.p. or more. In 1976 the production had increased to 10,511 with 5,305 at less than 200 h.p. and 5,206 over 200 h.p. These tractors are being produced in sizes of 400 h.p. and more. These tractors are especially popular in the grain producing areas.

Trend to Diesel Tractors

Gasoline tractors produced outnumbered diesels by 84,102 to 80,920 in 1961 (Table 16-4). This relationship had changed dramatically by 1976 when 177,034 diesel tractors were produced in the United States compared to only 16,690 gasoline and LPG tractors.

Safety Trends

Modern tractors are being designed and built to meet more specific safety standards. These safety features include

TABLE 16-3.   RETAIL SALES OF FARM WHEEL TRACTORS BY HORSEPOWER (IN UNITS)

1970 — 1972 — 1974 — 1976

Maximum Observed PTO Horsepower Classes

	9-34	35-39	40-49	50-59	60-69	70-79	80-89	90-99	100-109	110-119	120-129	130-139	140 & Over	TOTAL
1970	7,847	23,974	5,400	17,635	18,811	7,809	5,309	23,312	11,545	4,595	3,893	4,077	1,325	135,532

DATA: Farm & Industrial Equipment Institute.
* Figures do not include garden tractors, motor tillers or contractors' off-highway types. Report includes units made or imported by these ten firms: Allis-Chalmers; David Brown; Case; Deere; Deutz; Ford; International Harvester; Massey-Ferguson; Minneapolis-Moline and Oliver.
‡ Sales reported as miscellaneous are retail sales which cannot be classified by state or as U.S. Government, but can be reported by product, type and size. They include sales and dispositions that do not go through usual marketing channels such as gifts to non-profit organizations, intra-plant or company transfers etc.

Maximum Observed PTO Horsepower Classes

	Under 35	35-39	40-49	50-59	60-69	70-79	80-89	90-99	100-109	110-119	120-129	130-139	140 & Over	Total
1972	11,790	21,382	8,901	16,464	18,795	6,143	4,731	22,600	6,468	17,723	5,936	9,638	6,191	156,762

DATA: Farm & Industrial Equipment Institute. Figures do not include garden tractors, motor tillers or contractors' off-highway types. Report includes units made or imported by these 13 firms: Allis-Chalmers; David Brown; J. I. Case; Deere; Deutz; Ford; Franklin; International Harvester; Massey-Ferguson; Minneapolis-Moline; Oliver; Steiger; and Versatile. Sales reported as miscellaneous are retail sales which cannot be classified by state or as U.S. Government, but can be reported by product type and size. They include sales and dispositions that do not go through usual marketing channels such as gifts to non-profit organizations, intra-plant or company transfers, etc.

(continued)

TABLE 16-3. Continued

Maximum Observed PTO Horsepower Classes

	Under 35	35-39	40-49	50-59	60-69	70-79	80-89	90-99	100-109	110-119	120-129	130-139	140 & Over	4-WH Drive To 169	4-WH Drive 170 & Over	Total
1974	9,764	22,246	10,851	15,824	16,424	7,317	9,317	6,004	15,994	3,254	27,979	3,589	16,951	1,473	6,814	173,801

DATA: Farm & Industrial Equipment Institute. Figures do not include garden tractors, motor tillers or contractor's off-highway types. Report includes units made or imported by these firms: Allis-Chalmers; Case; Deere; Deutz; Ford; Franklin; International Harvester; Long; Massey-Ferguson; Steiger; Versatile; White; and distributors for Leyland and Muir-Hill tractors. Sales reported as miscellaneous are retail sales which cannot be classified by state or as U.S. Government, but can be reported by product type and size. They include sales and dispositions that do not go through usual marketing channels such as gifts to nonprofit organizations, intra-plant or company transfers, etc.

Two-Wheel-Drive Tractors by PTO Horsepower / Four-Wheel Drive Tractors by Net Engine Horsepower

	Under 35	35-39	40-49	50-59	60-69	70-79	80-89	90-99	100-119	120-129	130-139	140-149	150-159	160 & Over	Total 2-Wheel Drive	Under 200	Over 200	Total 4-Wheel Drive	All Farm Wheel Tractors
1976	8,410	7,499	16,012	14,140	15,810	8,446	9,857	1,201	16,595	20,031	6,456	5,127	7,970	5,124	142,678	5,305	5,206	10,511	153,189

DATA: Farm & Industrial Equipment Institute. Figures do not include garden tractors, motor tillers or contractor's off-highway types. Report includes units made or imported by these firms: Allied Farm Equip. (some Leyland and Satoh); Allis-Chalmers; J I Case; Deere; Deutz; Ford; International Harvester; Long; Massey-Ferguson; Steiger; Versatile; R. M. Wade & Co. (some Leyland and Satoh); and White. Sales reported as miscellaneous are retail sales which cannot be classified by state or as U.S. Government, but can be reported by product type and size. They include sales and dispositions that do not go through usual marketing channels such as gifts to non-profit organizations, intra-plant or company transfers, etc.

(Courtesy Implement and Tractor Intertec Publishing Corporation)

TABLE 16-4
PRODUCTION OF WHEEL TRACTORS (IN UNITS)
BY TYPE OF FUEL USED (EXCEPT GARDEN AND
CONTRACTORS' OFF-HIGHWAY TYPES)

Year	Gasoline	Diesel	LPG	TOTAL
1955	277,105	41.506	11,530	330,141
1956	177,077	26,762	10,815	214,654
1957	179,142	37,352	12,556	229,050
1958	173,040	55,864	12,365	241,269
1959	167,961	79,548	12,407	259,916
1960	84,306	62,033	5,848	152,187
1961	84,102	80,920	6,395	171,417
1962	99,414	80,293	8,394	188,101
1963	96,108	98,609	8,732	203,449
1964	90,004	115,439	7,778	213,221
1965	98,970	138,247	6,833	244,050
1966	107,683	157,352	5,652	270,687
1967	80,630	158,741	2,844	242,215
1968	61,978	147,744	3,477	213,199
1969	59,111	135,118	1,475	195,704
1970	48,981	122,622	(D)	171,603*
1971	40,642	126,859	(D)	167,501*
1972	42,389	154,827	(D)	197,216*
1973	37,860	173,646	(1)	211,506
1974	33,860	176,214	(2)	210,074
1975	27,429	183,484	(2)	210,913
1976	16,690	177,034	(2)	193,724

(D)—Withheld to avoid disclosing the operations of individual companies.
* Production totals exclude LPG powered tractors which amount to less than 1% of wheel tractors produced.
(1)—For 1973, gasoline and LPG powered tractors 90 horsepower and over have been included with diesel tractors to avoid disclosure of individual company data. These tractors amount to less than 1% of wheel tractors produced.
(2)—Since 1974, LPG and gasoline fuel type tractors are combined.
Data: Current Industrial Reports, Series M35S, Bureau of the Census.
(*Courtesy* Implement and Tractor Intertec Publishing Corporation)

roll-bars, safer cabs, and other safety features.

Roll bars are provided to reduce danger to the operator in case of rollover. Safety features provided in modern tractor cabs include reduced noise levels, cleaner air, improved visibility, safer seats, seat belts, and safer means of entering and leaving the cab. Wider tread settings are used for operation on hillsides. Oscillating wheels on four wheel drive tractors also provide improved safety on hillsides. Power steering, power brakes, and differential locking systems also provide added safety. Improved lighting and slow-moving vehicle emblems reduce hazards when operating tractors on roadways.

Operator Comfort

Farm tractors are now being built to provide for much greater levels of operator comfort. These features include power steering, comfortable cabs with heating and cooling (Figure 16-18), comfortable seats, and improved controls. Tractors are also now available with radio and stereo equipment. These comfort features also tend to lessen operator fatigue and improve safety.

SUMMARY

Tractor types include both wheel and track-type tractors. Wheel types include utility size tractors, large two-wheel drive tractors, and four-wheel drive tractors. Very few tricycle type tractors are currently manufactured. Four-wheel tractors are equipped with adjustable wheel treads for row crop use. Specially designed tractors are available for orchard, vineyard, lawn and garden, industrial, and other specialty uses.

The number of tractors sold and the number of tractors on farms are decreasing somewhat, but this is offset by a substantial increase in tractor size. A very high proportion of tractors being produced have diesel engines. There is a trend toward improved safety features and greater operator comfort in modern farm tractors.

FIG. 16-18.   Interior of a modern tractor cab. (*Courtesy* Deere and Company)

Shop Projects

A. Field trip to tractor dealership to view and discuss tractor types available

B. Examine current sales literature to determine types of tractors and options available

Questions

1. Which tractor types are most adaptable to your community? Why?

2. What are some of the special options and adaptations available on modern tractors?

3. What is happening to both the number of tractors sold and to the number of tractors on farms in recent years?

4. What is the current trend in tractor size?

5. What has been done to make modern tractors safer?

6. What are the trends that may be observed in providing greater operator comfort?

References

Current manufacturer's sales literature

17   Clutches, Transmissions, Differentials, and Final Drives of Farm Tractors

The engine is the source of power for the tractor and must be properly coupled to the rear-drive wheels to make the tractor a practical machine. It must be possible to disconnect the engine from the rear-drive wheels, or load, so that the forward or backward motion of the tractor can be started or stopped while the engine is running. When shifting gears on most tractors, it is also necessary to disconnect the engine from the rear-drive wheels. A *clutch* is the mechanism used to disconnect the engine from its load.

A speed-reducing system is placed between the engine clutch and rear-drive wheels. This is necessary because the engine must run at a relatively high speed to develop enough power to do its work, while the rear-drive wheels must run at a lower speed.

The transmission is used to select the proper gear ratio or speed for the job to be done. It also provides a reverse gear for moving the tractor backwards. Some speed reduction also takes place at the pinion and ring gear around the differential and at the final drive unit. The purpose of the differential is to equalize the power to the drive wheels when the tractor is turning. The final drive is another speed reduction unit that is used between the differential and the rear-drive wheels on many tractors. The relative location of these parts for a two

wheel drive tractor is shown in Figure 17-1. The clutch, transmission, differential, and final drive of a tractor form its power train.

CLUTCHES

The clutch is directly in back of the engine flywheel. The flywheel is often a part of the clutch assembly. Most tractor clutches are of the disc type, having one or more driving plates. A common clutch is the single-driving-plate type (Figure 17-2).

The driving plate (1) is lined with friction surfaces that are pressed against the driven plates (2) and (3) when the clutch is engaged. When the clutch is disengaged, the parts are separated and the driving plate turns with the engine while the driven plates remain stationary. The operation of this and similar clutches is shown in Figure 17-3. When the pedal is down, the driving disc (disk) is disengaged from the driven plate and flywheel. The driven shaft does not turn. When the clutch pedal is up, the clutch is engaged. The clutch disc (disk) and driven plate are held in contact with the flywheel by pressure from the clutch spring. The parts revolve as a unit and the driven shaft delivers power to the transmission. This is known as a dry

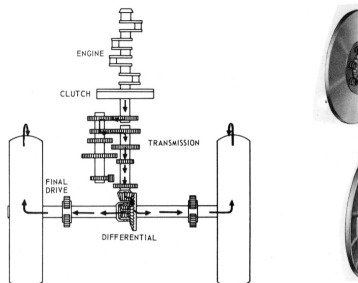

FIG. 17-1. The power train of a farm tractor: clutch, transmission, differential, and final drive. (*Courtesy* Deere and Company)

FIG. 17-2. An "exploded" view of a clutch with a single driving plate. When the driving clutch is engaged, the driven plates (2 and 3) are forced against the driving plate (1) by the yoke assembly (4). (*Courtesy* Minneapolis-Moline Division of Motec Industries, Inc.)

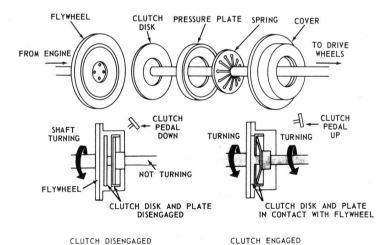

FIG. 17-3. Top of diagram shows the parts of a disc (disk) type clutch. The disengaged clutch is at the left and the engaged clutch at the right. (*Courtesy* Deere and Company)

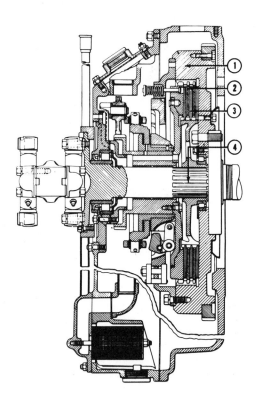

type of clutch because it does not run in oil.

The multiple-disc clutch (Figure 17-4) is commonly used as a steering clutch on crawler tractors. It is similar to the single-disc clutch, but has many driving discs and many driven discs. This type of clutch usually runs in oil and is known as a wet clutch. However, some multiple-disc clutches are also of the dry type.

Clutches must be adjusted so that the parts will run freely when the clutch is disengaged but will make good contact to avoid slippage when the clutch is engaged. A clutch that slips will get hot and wear rapidly. Some clutches are spring-loaded and the parts are held in firm contact with one another by spring pressure when the clutch is engaged. These are usually foot-operated clutches (Figure 17-5).

Some clutches are operated by a hand lever and cam mechanism. The

FIG. 17-4.   A multiple-disc clutch mounted in the engine flywheel: the flywheel (1), driving plate (2), driven plate (3), and clutch shaft (4). (*Courtesy* Caterpillar)

FIG. 17-5.   A spring-loaded foot-operated clutch. The spring pressure clamps the driven disc between the pressure plate and the fly wheel. The foot pedal is used to disengage the clutch. (*Courtesy* International Harvester Company)

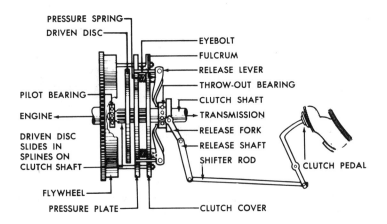

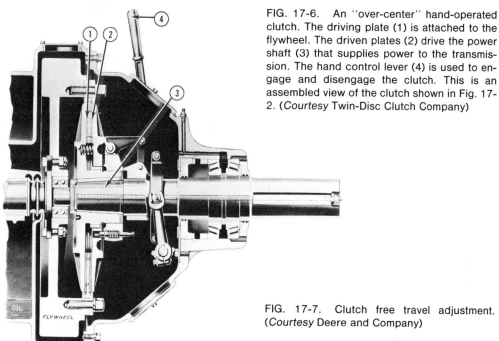

FIG. 17-6. An "over-center" hand-operated clutch. The driving plate (1) is attached to the flywheel. The driven plates (2) drive the power shaft (3) that supplies power to the transmission. The hand control lever (4) is used to engage and disengage the clutch. This is an assembled view of the clutch shown in Fig. 17-2. (*Courtesy* Twin-Disc Clutch Company)

FIG. 17-7. Clutch free travel adjustment. (*Courtesy* Deere and Company)

lever goes "over-center" when the clutch is engaged and will stay there until released (Figure 17-6).

Clutch Adjustments

The spring-loaded type of clutch normally has a clearance of about one-sixteenth of an inch between the clutch-release lever and the clutch-release bearing. This is amplified through the clutch pedal and its linkage so that a movement of about one to two inches is required at the clutch pedal to take up the clearance. This is called *free travel*. Normal clutch wear can usually be compensated for by adjusting the free travel. On most tractors, this adjustment is made in the clutch-pedal linkage (Figures 17-7 and 17-30).

The hand-operated, or over-center, type of clutch is adjusted by releasing a lock pin and turning the adjusting yoke in or out to get the desired adjustment. Turning the adjusting yoke in toward the clutch plates will compensate for wear (Figure 17-29). Since clutch and linkage designs vary with makes of tractors, it is advisable to consult the tractor operator's manual for specific recommendations on clutch adjustments.

TRANSMISSIONS

The transmisson is a speed-reducing mechanism equipped with several gear ratios. The operator can select a ratio to meet load and speed requirements. This is necessary because an engine must run near its rated speed to develop power near its rated capacity. The drive wheels of the tractor must turn at a relatively slow speed. The transmission also provides a reversing gear.

Simple transmissions consist of gear sets that make it possible to obtain the proper output shaft speeds. When two gears mesh with each other their speeds will be inversely proportional to the number of teeth in each gear. For example, when a 30 tooth gear meshes with and is driven by a 10 tooth gear, the 30 tooth gear will turn at one third of the speed (RPM) of the 10 tooth gear. The engineer that designs the transmission selects the proper gears to provide the transmission with the desired gear ratios and speeds.

The transmission fits into the power train as shown in Figure 17-1. The principle of a simple gear transmission is shown in Figures 17-8, 17-9 and 17-10. Figure 17-8 shows the transmission in low gear. This gear provides the greatest speed reduction. Low gear is used when high torque (pulling force) is required at the drive wheels to start and pull a heavy load at a low speed. Second gear, Figure 17-9, provides a faster field speed at a lower torque and drawbar pull than first gear will provide. Additional forward speeds are available through similar gear sets in tractor transmissions. The highest speed available is usually the "road gear" which will permit speeds up to about 18 or 20 miles per hour.

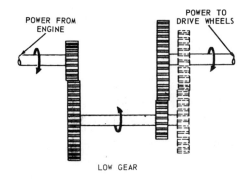

FIG. 17-8.   Transmission in low or first gear. (*Courtesy* Deere and Company)

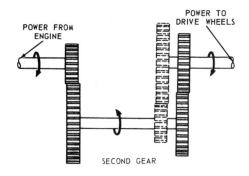

FIG. 17-9.   Transmission in second gear. (*Courtesy* Deere and Company)

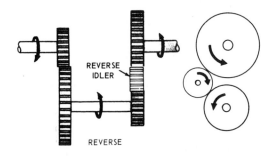

FIG. 17-10.   Transmission in reverse gear. Note reverse idler gear used to reverse direction of rotation of output shaft to drive wheels. (*Courtesy* Deere and Company)

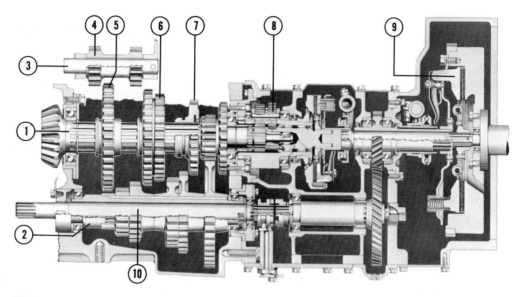

FIG. 17-11. Tractor transmission, clutch, independent PTO, and Torque-Amplifier: main shaft (1), countershaft (2), reverse idle shaft (3), reverse idler gear (4), first and reverse sliding gear (5), second and third sliding gear (6), fourth and fifth sliding gear (7), Torque-Amplifier unit (8), clutch (9), and independent PTO shaft (10). (*Courtesy* International Harvester Company)

Figure 17-10 shows how a reverse gear is provided in a transmission. When the reverse idler is in mesh with two other gears, as shown, then the output shaft, which delivers power to the rear wheels, is reversed in direction. One or two reverse speeds are provided in most farm tractors.

The sectional view in Figure 17-11 shows the gears, the main shaft, and the countershaft of a transmission. Many tractor transmissions provide for five or six forward and one or two reverse speeds. Some older tractors may have only one reverse and a few forward speeds, but some current model tractors have ten or more forward speeds and several reverse speeds.

When driving most tractors, it is not necessary to shift through several gears as is done with an automobile. The tractor gear is selected to fit the load and

speed conditions, and the forward motion of the tractor is started without first going through the lower gears. The gear is selected with the gearshift lever (Figure 17-12). This lever moves sliding gear sets in the transmission. When in gear, two gears on the countershaft mesh with one on both the drive shaft and the main shaft of the transmission.

The transmission described above is known as a sliding gear transmission because the shift lever is used to slide the gears into mesh as required for the proper tractor speed. Some transmissions, similar to the sliding gear transmission, have the gears in mesh at all times, but use a shift collar to engage the desired gear sets. These are known as collar shift transmissions.

Another transmission that keeps the gear sets in constant mesh is the synchromesh transmission (Figure 17-13). A

FIG. 17-12.   Gearshift lever mounted on the transmission. (*Courtesy* John Deere, Moline, Illinois)

FIG. 17-13.   A synchromesh transmission. (*Courtesy* Deere and Company)

synchronizer has been added that will bring the parts to be mated to the same speed before they engage. This results in smooth shifting. Manual shift synchro-mesh transmissions are used on many automobiles and other automotive units as well as on farm tractors.

The operation of a synchronizer unit is shown in Figure 17-14. The internal and external cones come together before the mating parts mesh. The spring loaded balls allow the friction

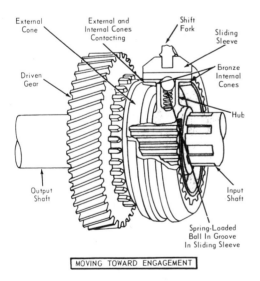

FIG. 17-14.   A synchronizer for a synchromesh transmission. (*Courtesy* Deere and Company)

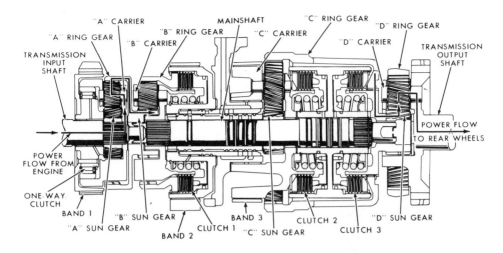

FIG. 17-15.   Cross-section view of a transmission with power shift. Gear-shifting is accomplished by several clutches and planetary gear units. The operator selects the desired gear ratio with a hand lever. (*Courtesy* Ford Motor Company)

cones to synchronize or equalize the speeds of the moving parts before meshing. Additional movement will compress the springs under the balls and will allow the internal teeth on the sliding sleeve to fully engage the external teeth on the side of the driven gear so they will run as a unit.

While most tractors are equipped with a conventional selective gear transmission, as described in the preceding paragraphs, there are some that provide features that make a tractor more versatile as a source of farm power. The power-shift transmission (Figure 17-15) makes the shifting of gears possible without stopping the tractor or using the clutch. It provides for ten forward speeds and two reverse speeds, as well as park and neutral positions. Shifting is accomplished on-the-go through hydraulically operated clutches and brakes that control three planetary-gear sets. The shifting lever can be moved through its complete range of speeds while the

tractor is in motion. Table 17-1 shows the speed ranges possible with this transmission.

Another device that is used by some tractor manufacturers permits reducing

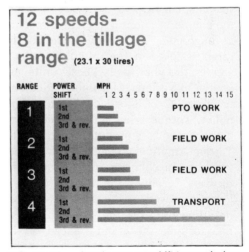

FIG. 17-16.   Another power shift transmission has four speed ranges with three power shift positions within each range. (*Courtesy* J. I. Case Company)

TABLE 17-1.   GROUND SPEEDS POSSIBLE WITH "POWER-SHIFT" TRANSMISSION IN MILES PER HOUR (12 x 28 REAR TIRES)

Gear	Engine rpm		
	1200	**1750**	**2200**
	GROUND SPEEDS POSSIBLE		
10	9.8	14.3	18.1
9	6.7	9.7	12.1
8	4.1	6.0	7.5
7	3.2	4.6	5.8
6	2.8	4.0	5.1
5	2.1	3.1	3.9
4	1.3	2.0	2.5
3	1.0	1.4	1.7
2	.9	1.3	1.6
1	.6	.9	1.2
Neutral	—	—	—
R1	1.9	2.8	3.5
R2	2.8	4.1	5.1
Park	—	—	—

(**Courtesy** Ford Motor Company)

the forward speed of tractors without reducing engine speed or stopping when the engine becomes overloaded because of tough going in the field. Reducing the forward speed of the tractor permits a greater drawbar pull. These units are known by various trade names. The unit in Figure 17-17 is called a Torque-Amplifier. Torque is an action which produces, or tends to produce, rotation. The Torque-Amplifier consists of a planetary-gear mechanism that reduces the forward speed of the tractor when a lever is moved to the proper position. Resetting this lever to its original position restores the original speed of the tractor. The range of speeds possible with the Torque-Amplifier is illustrated in Figure 17-18. Mechanisms of this type make two speeds available in each gear. This actually doubles the number of forward and reverse speeds in the transmission.

FIG. 17-17.   A Torque-Amplifier unit used with a standard transmission permits shifting down without using the clutch or stopping the tractor. It is a planetary gear unit. Fig. 17-11 shows the location of this unit in relation to the transmission and the clutch. (*Courtesy* International Harvester Company)

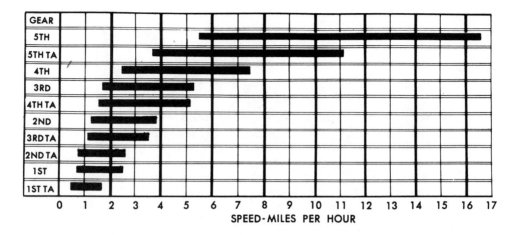

FIG. 17-18. The gear ratios of a regular transmission are doubled when a Torque-Amplifier unit is used. (*Courtesy* International Harvester Company)

HYDRAULIC TORQUE CONVERTERS

As mentioned before, torque is an action which produces, or tends to produce, rotation. Assume that a tractor engine has a certain torque at its rated speed. Because this torque is too low to start a heavy load at the drawbar, you shift the tractor transmission into low gear. The engine is now turning at rated rpm with low torque, but the rear wheels of the tractor are turning at a low rpm and have high torque. In effect, the low engine torque has been converted to high torque at the drive wheels. A mechanical gear transmission is a form of torque converter. However, the name is more commonly applied to a hydraulic device that has come into use on farm tractors. One manufacturer* describes the action of this unit (Figure 17-19) as follows:

---

* J. I. Case Company, Racine, Wisconsin.

"In a farm tractor equipped with a hydraulic torque converter, the engine is the primary source of power. The engine flywheel (A) is connected to the impeller (B), which is ringed with pitched blades. As the flywheel turns, it turns the impeller, which pumps the converter oil. The driven oil strikes the blades of the turbine (C), forcing it to spin, creating input torque (force). The swirling oil rebounds off the turbine blades against the stator (D), which is anchored. The flow of oil against the stator creates a "backing-up" force, similar to the force created by squirting a high pressure hose against a wall. This is reaction torque and it is added to the force of the input torque.

The swirling oil bounces off the stator and is redirected back into the spinning impeller blades, taking some of the work load off the engine. Thus, the circulating oil is driven around and around, bouncing from impeller to turbine to stator to impeller, helping itself to multiply torque.

Note that the turbine is connected to the converter output shaft which turns with the turbine; its twisting power (torque) is geared to the rear wheels of the tractor."

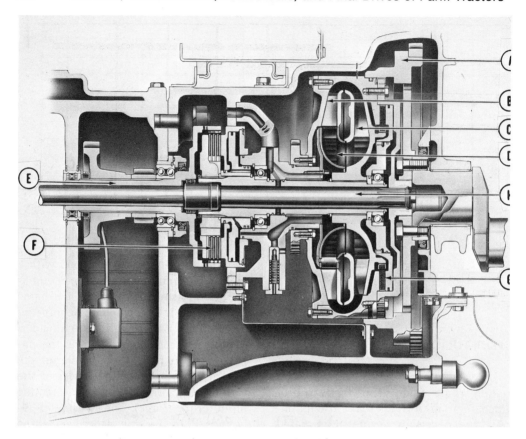

FIG. 17-19.   A hydraulic torque-converter takes low torque at high speeds and converts it to high torque at low speeds: flywheel (A), impeller (B), turbine (C), stator (D), converter output shaft (E), multiple-disc clutch (F), single-disc clutch (G), PTO shaft (H). (*Courtesy* J. I. Case Company)

The hydraulic torque converter drives through the fluid in the unit. Some slippage results, especially at low engine speeds. When a tractor equipped with a torque converter encounters a heavy load, the forward speed of the tractor is decreased. As the load becomes lighter, the forward speed is increased. The tractor ground speed is actually adjusted to meet load conditions. The engine rpm during these changes in load and forward speed remains nearly constant. The characteristics of a torque converter are:

1. The forward speed of the tractor automatically adjusts to the load at the drawbar in each transmission gear drive.
2. Less shifting of gears is required.
3. There is less wear on the tractor power train because of the fluid connection between the engine and the transmission.
4. Slippage in the torque converter results in slightly higher fuel consumption than when a direct gear drive is used.

THE DIFFERENTIAL

When a tractor is driven in a straight line, both rear wheels turn at the same speed. However, when it is driven on a turn, the outside wheel must travel farther than the inside wheel; therefore, it must turn faster. The differential in the tractor power train (Figure 17-1) permits the two rear wheels to turn at different speeds while power is being transmitted through both wheels. The diagram in Figure 17-20 shows the principle of the differential. The ring gear is driven by the bevel pinion, which receives its power from the transmission through the spline shaft.

The four differential bevel pinions are mounted in a housing on steel shafts. This housing is attached to the ring gear and turns with it. The bevel pinions mesh with the two side gears. The side gears are attached to the two countershafts. The countershafts transmit power to the rear-drive wheels. When the tractor is moving straight forward, the entire differential assembly turns as a unit and the side gears do not turn relative to the differential bevel pinions. When making a turn, one countershaft turns faster than the other, causing the side gears to turn in relation to the differential bevel pinions. Applying the brakes on one wheel will also cause the side gears to turn relative to the differential bevel pinions. In fact, one wheel can be stopped and all of the power will be transmitted to the other wheel. If one wheel is stopped, the other wheel turns at twice its original speed. This causes the tractor to turn rather rapidly, which must be considered when making a turn with the brake applied on one wheel.

When one wheel of a tractor is on a slippery surface while the other wheel is on firm ground, the wheel that is on the slippery surface will spin because of the construction of the differential. The wheel that is on firm ground will not turn. All of the power is being applied to the wheel that spins.

The principle of the differential lock as used on some tractors is shown in Figure 17-21. The sliding collar "C" locks

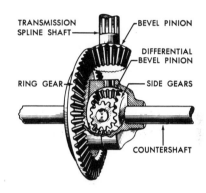

FIG. 17-20. Cutaway view of differential gears. (*Courtesy* International Harvester Company)

FIG. 17-21. Differential lock. (*Top*) Both rear-axle shafts turning at the same speed. (*Center*) The wheel on axle B turns faster than that on A because of uneven traction. (*Bottom*) Slight heel pressure on the differential lock pedal causes collar C to slip into place, locking the two half-shafts together. Equal power is then delivered to both rear wheels. (*Courtesy* International Harvester Company)

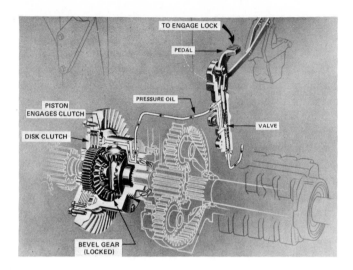

FIG. 17-22.   A hydraulically operated differential lock. (*Courtesy* Deere and Company)

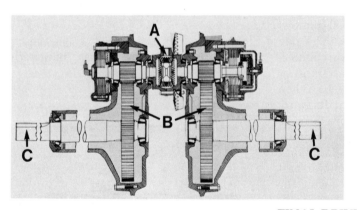

FIG. 17-23.   Final-drive gears located in the differential housing. (*Courtesy* International Harvester Company)

the two halves of the rear axle together so that equal power is supplied to both rear wheels. The device can be engaged, mechanically or hydraulically, by the operator when needed. Figure 17-22 shows a hydraulically operated differential lock. The differential functions normally when the lock is not engaged.

FINAL DRIVE

The last speed reduction unit is usually a set of final drive gears located between the differential and the drive wheels of the tractor. It is used on most farm tractors, although there are some drives that accomplish sufficient speed reduction in the transmission and at the bevel and ring gears to make the final drive gears unnecessary. The final drive gears may be located in the housing with the transmission and differential (Figure 17-23). However, they are more com-

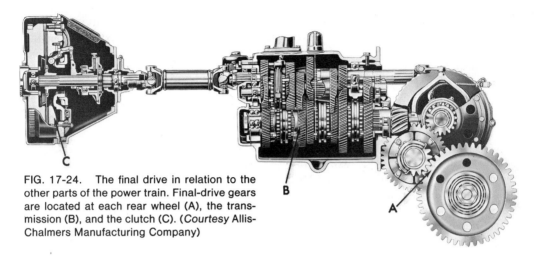

FIG. 17-24.   The final drive in relation to the other parts of the power train. Final-drive gears are located at each rear wheel (A), the transmission (B), and the clutch (C). (*Courtesy* Allis-Chalmers Manufacturing Company)

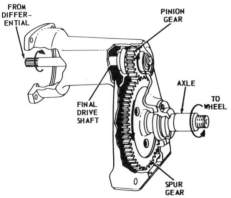

FIG. 17-25.   Pinion and spur gear set forms a speed reduction in the final drive. (*Courtesy* Deere and Company)

monly located near each drive wheel, either in the form of a pinion and spur gear set (Figures 17-24 and 17-25) or in the form of a planetary gear set as shown in Figure 17-26.

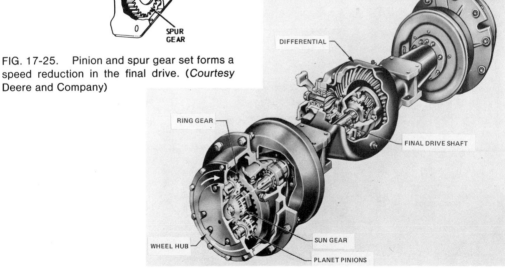

FIG. 17-26.   Final drive using planetary gear speed reduction units near each drive wheel. (*Courtesy* Deere and Company)

FOUR WHEEL DRIVE TRACTORS

As farms have increased in size, a demand for large tractors has developed. Large farm tractors, those over about 150 H.P. in size, usually use a four wheel drive chassis. The four wheel drive system provides better traction characteristics and permits better utilization of the engine's power for field work than would be possible with a two wheel drive tractor of the same size. Usually there is less wheel slippage when a four wheel drive unit is used. Dual tires can be added to both two and four wheel drive units to add more soil contact area if it is needed.

The power train of a four wheel drive tractor is shown in Figure 17-27. The flow of power is from the engine to the transmission, upper drive shaft, drop box, lower drive shafts, differentials, axles, and then to the drive wheels. The power from the engine is divided at the drop box. Drive shafts extend from the drop box to the front and rear differentials. The differentials act in the same manner as they do on two wheel drive tractors.

SERVICING THE TRANSMISSION, THE DIFFERENTIAL, AND THE FINAL DRIVE

The transmission, the differential, and the final drive must be designed to take the shock loads encountered in everyday tractor use. To insure long life of these parts they must be lubricated properly and serviced regularly.

In many tractors, these parts are all in the same housing and are lubricated by a transmission oil that partly fills this housing. The oil is carried to all moving parts and bearings by the rotation of the

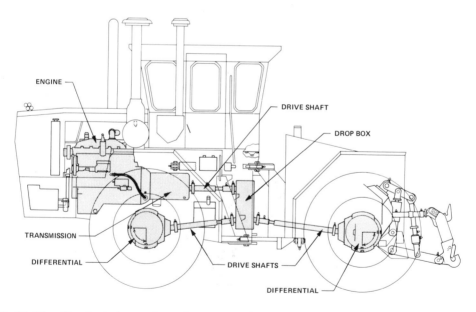

FIG. 17-27.   The shaded area shows the power train of a four wheel drive tractor. (*Courtesy* Steiger Tractor, Incorporated)

larger gears and chains which dip into the oil supply. This should be kept at its proper level to insure good lubrication.

The lubricant in the transmission gear case may become contaminated with dust, condensation (water), and chips of metal that wear from gears and other parts. For this reason, the oil in the transmission should be changed periodically, following the manufacturer's recommendations.

Some tractor manufacturers recommend SAE 90 transmission oil for summer use, and SAE 80 for winter use when temperatures drop well below freezing. The reason for this is because a summer lubricant may become too viscous to do a good lubricating job during extremely cold weather use. Since lubrication recommendations vary with makes and models of equipment, it is important to follow the manufacturer's recommendations in selecting lubricants for such vital parts as the transmission, the differential, and the final drive units.

RUBBER TIRES

Rubber tires have been standard equipment on farm tractors for many years. They have made it possible to increase the speed of field operations, increase the efficiency of the tractor at the drawbar, and provide greater comfort for the operator. In order to get the best performance from tractor tires, it is helpful to understand their characteristics.

Inflation Pressures

It is very important to use proper inflation pressure in order to get satisfactory performance and wear from tractor tires. Air in the tire carries the load, therefore, the pressure should be adjusted to the load on the tire.

Normal inflation pressures for rear wheel tractor tires vary between 12 and 40 lbs. per square inch. In general, as inflation pressures are decreased, traction is improved. However, tire life will be shortened by underinflation as well as overinflation.

Underinflation will cause abnormal flexing of the tire walls and this repeated buckling of the side wall will result in a series of breaks and separation of the cord fabric. This flexing action will also result in side wall cracks. The minimum recommended inflation pressure for most rear wheel tractor tires is 16 lbs. per square inch. If the tire is filled with a solution the pressure should be determined with the valve down. A liquid type of pressure gauge should be used and the gauge should be washed after each use. If it is not practical to determine the tire pressure with the valve down, it may be checked with the valve up, but it will be necessary to add about one half pound to the pressure gauge reading for each foot height of liquid in the tire in order to get the actual pressure in the tire.

Overinflation should be avoided because, under most farm conditions, the traction qualities of the tire are not as good when the tire is inflated to higher pressures. This will result in more slippage and greater tire wear. The tire is also more susceptible to damage by stones and other sharp obstructions. Maximum inflation pressures vary with the tire size and load; however, few farm operations require rear tire pressure over 20 pounds per square inch. Tires used on hard surfaces such as highways will wear longer when slightly overinflated.

On two wheel drive tractors, the front tires are usually inflated to pressures between 24 and 36 pounds per square inch. The recommended pressure again will depend upon the tire size and load. Follow manufacturer's recommendation as stated in the tractor owner's manual.

Inflation pressures should be checked every two or three weeks and should be maintained at the recommended level for the load being carried.

Adding Weight to Rubber Tires

Most tractors, as they come from the factory, do not have enough weight at the drive wheels to give the best performance. The maximum drawbar pull of a tractor equipped with rubber tires can be increased and slippage can be decreased when extra weight is added to the drive wheels. Weight can be added with liquid ballast inside the tire, cast iron on the wheel, or a combination of these.

Cast-iron weights are available from the tractor manufacturer. Liquid weight is added in the form of water in areas where the temperature stays above freezing. A solution of water and calcium chloride is recommended where freezing temperatures occur. Tire dealers have the necessary equipment for mixing and installing calcium chloride solutions for ballast.

Dual Tires

Dual tires are used on many farm tractors. The increase in engine horsepower on tractors has brought about some traction problems. The addition of dual tires permits more weight to be added to the tractor and improves its traction. Dual tires give more ground contact which results in less unit pressure on the soil and more floatation. Tractor stability is increased by the increase in width of the tractor when duals are added. Dual tires also permit the tractor to be used under more adverse conditions thus permitting earlier field work in the spring.

Dual tires are not recommended for all tractors. Be sure that the tractor axle is designed to take the heavier loads and resulting stresses that will be imposed on it when duals are used. Most farm operations can be completed without the use of duals. This is especially true under normal dry field conditions with drawbar loads that are not excessive.

Wheel Slippage With Rubber Tires

Rubber tires will perform best when the wheel slippage is between 10 and 15 percent. Drive wheel slippage can be checked easily by driving a tractor with no load at the drawbar a measured distance of 100 feet. Count the number of revolutions to go this distance. Now drive the same tractor with its drawbar load over the same route for the same number of revolutions and put a mark at that point. The difference in the distance that the tractor travelled with no load and with a load at the drawbar is the percent of slippage. For example, if the tractor made 6.5 revolutions in going the distance of 100 feet with no load, but with a load at the drawbar it travelled only 88 feet in the same number of revolutions (6.5) then its slippage is 12 percent. Excessive slippage will result in rapid tire wear.

TRACKS

Crawler tractors were developed to operate in peat land areas, but have become popular for use in some farming

FIG. 17-28.   A crawler tractor pulling three 14-foot grain drills. (Photo by Authors)

operations. They have good traction over a wide range of conditions and are very maneuverable. However, crawlers are usually limited to speeds of less than six miles per hour.

Crawler tracks are available with several kinds of shoes. These shoes make the crawler suitable for a variety of work. Many farmers use crawler tractors on large acreages where several machines are pulled behind one tractor (Figure 17-28). They are also used in hilly areas because they are quite stable on hillsides and on rough land.

SUMMARY

The clutch, transmission, differential, and final drive are the parts that make the tractor engine a versatile farm power unit. These parts provide ways of starting, stopping, and changing the speed and drawbar pull of the tractor. They adapt the tractor engine to the job it is to perform.

The clutch is used to disconnect the engine from its load. Clutches commonly used in tractors are the single-driving-plate and multiple-disc types. Clutches can be hand- or foot-operated. Clutches must be adjusted so that the parts will run freely when disengaged and will not slip when engaged.

The transmission is a set of gears that provides several speed-reduction ratios between the engine and the final drive. This is necessary because the engine must run at a relatively high speed to develop maximum power and the drive wheels must run at a much slower speed. The transmission also provides a reverse gear.

Hydraulic torque converters are used on some tractors. These permit the forward speed of the tractor to vary with the load at the drawbar which reduces gear shifting. The fluid coupling in the torque converter eases shock loads on the tractor power train, but slippage in the torque converter results in slightly higher fuel consumption.

The differential permits the drive wheels to turn at different speeds while the tractor is making a turn. Power is delivered to both drive wheels.

Most tractors are equipped with a final gear-reduction unit called the final drive. It is usually located between the differential and the drive wheels.

All these parts are enclosed and may easily be forgotten, but this must not be allowed to happen because expensive repairs will be needed if maintenance is neglected.

The tractor power-train units must be serviced and maintained properly to insure long, trouble-free service. The manufacturer's recommendations should be followed when servicing these units. Keep the operator's manual in a place where it is readily accessible.

Good traction and long life of tires are insured when the tires are properly inflated and weighted. Excessive slippage results when tires are overinflated and not properly weighted. This causes short tire life. Underinflation also results in short tire life because of excessive tire buckling.

Shop Projects

A.   Adjusting the clutch

To adjust a *hand-operated clutch* on a tractor, it is usually necessary to remove a side plate, pull out the lock pin, tighten the clutch, reset the lock pin, and replace the side plate (Figure 17-29).

It is best to follow the instructions in the operator's manual for the exact clutch adjustment for the tractor selected. These vary from one make or model of tractor to another.

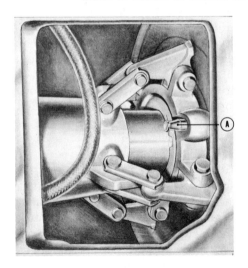

FIG. 17-29. A hand-operated over-center clutch is adjusted by first disengaging the clutch, then pulling out the lockpin (A) and rotating the clutch yoke assembly to the right until the lockpin slips into the next hole. The clutch lever should engage the clutch with a snap. (*Courtesy* Minneapolis-Moline Division of Motec Industries, Inc.)

The adjustment on a *foot-operated clutch* is usually one of adjusting the clutch linkage to get the correct amount of free travel (Figure 17-30).

Again we suggest referring to the tractor operator's manual for the exact instructions on making this adjustment on the tractor you have selected, because the instructions vary from one make or model of tractor to another. (Also see Figure 17-7.)

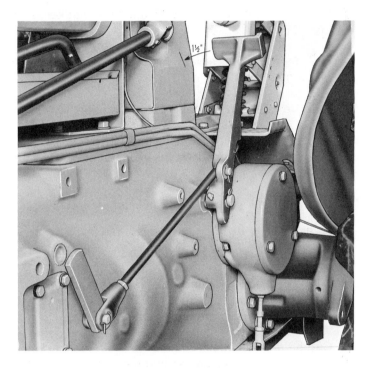

FIG. 17-30. A foot-operated clutch must have some free travel when properly adjusted. In this case, it is 1½ inches. The locknut is at (A) and the adjusting clevis at (B). (*Courtesy* Minneapolis-Moline Division of Motec Industries, Inc.)

B. Changing the transmission oil

Most tractor manufacturers recommend changing the oil in the transmission and differential case at least once a year or every 1000 hours. The procedure may vary with some tractors, but the following general instructions should be followed:

1. Warm up the tractor before draining the oil. At the end of a day's work is an excellent time to change the oil.

2. Remove the transmission drain plug (or plugs) at the bottom of the transmission and differential case and allow the oil to drain out. Do not confuse the hydraulic system drain plug with the transmission drain plug. See the operator's manual.

3. Replace the drain plug and fill the case with kerosene or furnace oil.

4. Run the tractor for several minutes to wash the transmission and differential parts.

5. Remove the drain plug again to drain out all flushing oil.

6. Replace the drain plug and fill the transmission to the proper level with oil recommended by the manufacturer. See the operator's manual.

C. Making a cutaway model of the transmission and differential

1. Obtain an old automobile or a light truck transmission and differential from a wrecked vehicle or from a used auto parts dealer.

2. With a hacksaw, cut the side of the transmission housing away so that the gears will be exposed. It may be necessary to disassemble the transmission to make cutting easier.

3. Cut the differential housing in the same manner and also cut the axles and axle housings so that the overall width of the unit is reduced to about 30 inches.

   Some welding will be necessary to replace the axle ends and axle-housing ends so that the brake drums can be left on the unit (Figure 17-31).

4. Assemble the unit as shown and weld a crank on to the front end of the transmission. Note that the transmission is laid on its side so that the gears are more readily visible.

5. Build a stand as shown. Paint all parts. Use contrasting colors.

6. The unit can now be used to demonstrate the principles of the transmission and differential.

FIG. 17-31.   This automobile transmission and differential cutaway shows the working parts. They were mounted on a portable stand. (Photo by Authors)

Questions

1.   What is the purpose of a clutch? Where is it located in the power train? Why?

2.   What is a dry clutch? A wet clutch?

3.   Where can you find specific instructions for adjusting a tractor clutch?

4.   Why are several forward speeds necessary in a tractor transmission?

5.   What is torque? What is a torque booster?

6.   Why is a differential used in the power train on most tractors? How does it function?

7.   Some tractors do not have a differential. Identify this type of tractor and explain why it does not need a differential.

8.   What is the final drive? Is it used on all tractors?

9.   How often should the oil be changed in the transmission and differential housing?

10. Where can you find recommendations on the type of oil to use in the transmission and differential?

11. What is the recommended air pressure for rear tires on a tractor? Should this pressure ever be changed? Why?

12. How can weight be added to rubber tires for tractors? Why is it necessary to use added weight?

13. What are the advantages of tracks as used on crawler tractors? The disadvantages?

14. Why is it important to keep the tractor operator's manual in a place where it can be found when needed? Suggest a good place to keep it.

References

Commercial Literature, J. I. Case Company, Racine, Wisconsin.

Fundamentals of Machine Operation, Preventive Maintenance, Deere & Company, Moline, Illinois, 1973.

Fundamentals of Machine Operation, Tractors, Deere & Company, Moline, Illinois, 1974.

Fundamentals of Service, Power Trains, 3rd Edition, Deere & Company, Moline, Illinois, 1977.

Tractor Maintenance, American Association for Vocational Instructional Materials, Athens, Georgia, 1975.

Tractor Transmissions, American Association for Vocational Instructional Materials, Athens, Georgia.

18 PTO Shaft, Steering Gear, Brakes, and Belt Drives

A farm tractor becomes a versatile power unit that will do a wide variety of farm jobs when it is equipped with a power take-off shaft. A tractor with this equipment can be used to deliver power to stationary machines as well as to field machines that require rotating power.

Whenever a tractor is used to drive a machine with a power take-off shaft, there is always a question of operating speed. This chapter will discuss power take-offs, and operating speed, as well as the tractor steering mechanisms, brakes, and belt drives.

POWER TAKE-OFF SHAFTS

Most farm tractors are equipped with a power shaft that extends from the rear of the tractor to a point above the drawbar. This shaft is known as the power take-off (PTO) shaft. The PTO shaft is used to drive such field machines as hay balers, corn pickers, and swathers and is also used in a stationary position to operate other machines, such as self-unloading forage wagons and irrigation pumps. On many older tractors, the PTO shaft stops when the clutch is disengaged to stop the forward motion of the tractor. In many cases, this is an inconvenient arrangement because it is often desirable to keep the machine running without interruption when the

tractor is stopped. Most recent-model tractors are equipped with a live or continuously running PTO shaft. Stopping the forward motion of the tractor does not stop the PTO shaft. This requires an additional clutch in the power train. One clutch is used to stop and start the movement of the tractor, while the second is used to control the PTO shaft.

The double clutch arrangement shown in Figure 18-2 requires only one control pedal. Depressing the foot pedal part way stops the forward (or backward) motion of the tractor. Depressing the pedal all the way also stops the PTO shaft. Other control systems use a separate control pedal or lever for each clutch.

Some farm operations, such as planting and fertilizing, require that the PTO shaft speed be proportional to the ground speed of the tractor. Some PTO shafts are equipped with a ground drive as well as an engine drive. This is motivated by a drive from the differential pinion shaft. The shifting of a lever makes it possible to select either type of drive (Figure 18-3).

In order for the PTO shaft to be most useful, it must have a standard speed and a standard shaft size. There are two standards in use at the present time. The old standard speed, which is found on most farm tractors made before 1959, is 540 ± 10 rpm. The shaft has

FIG. 18-1.   Power take-off (PTO) shaft with shield for safety. (Photo by Authors)

six splines and is 1⅜ inches in diameter. Some small tractors may be equipped with 1⅛-inch shafts and some larger tractors use a 1¾-inch shaft. The standard established in 1958 calls for a speed of 1000 ± 25 rpm with a 1⅜-inch shaft having 21 splines. The higher shaft speed permits the transmitting of more power with the same size shaft than was transmitted with the old standard PTO. On some tractors, it is possible to obtain both the 540- and the 1000-rpm speeds by a simple shift of a lever. Two PTO spline shafts are provided so that the tractor can be used with machines designed for either PTO speed.

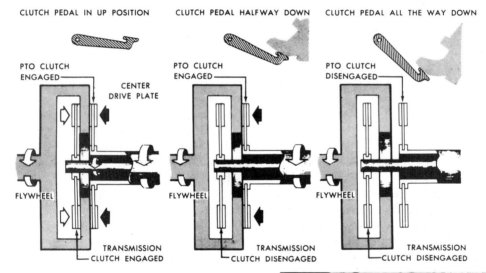

FIG. 18-2.   A double clutch that permits stopping the forward motion of the tractor without stopping the PTO shaft. (Drawing by Roger Cossette adapted from illustrations by the Ford Motor Company and the Long Manufacturing Division of Borg-Warner Corporation)

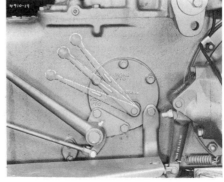

FIG. 18-3.   The PTO shaft on this tractor can be driven from the ground or from the engine, as indicated by the control lever. (*Courtesy* Massey-Ferguson, Limited)

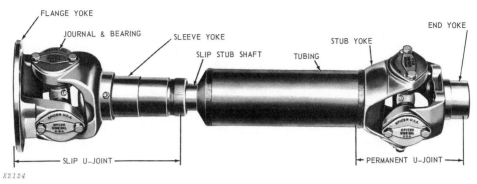

FIG. 18-4. A telescoping PTO shaft showing universal joints (U-joints) and telescoping feature of shaft (slip joint), (*Courtesy* Deere and Company)

A telescoping shaft with universal joints is used to transmit power from the tractor PTO shaft to the driven machine (Figure 18-4).

Metal safety shields are provided to cover the telescoping shaft and universal joints. These shields should always be in place when a PTO-driven machine is in use.

Some PTO safety shields are of the "spinner" type. The spinner safety shield rotates with the shaft except when contact is made with the spinner shield. It will then stop and permit the PTO shaft to rotate inside of the shield. Bell housings cover the universal joints. PTO shields are standardized so that one manufacturer's machine can be attached to another manufacturer's tractor.

STEERING MECHANISMS

Wheel Tractors

Tractor designs vary, as do steering mechanisms. There are some basic parts that are common to many systems. Most steering wheels need to be rotated about two revolutions to turn the tractor from straight forward to full right or full left. A gear-reduction unit is provided on all wheel tractor steering systems. The worm-and-lever mechanism (Figure 18-5) and the worm-and-sector unit (Figure 18-6) are in common use. They provide the necessary gear reduction to make steering easy under most conditions. Both types of units are used on both three- and four-wheel tractors.

FIG. 18-5. (A) Worm- and (B) lever-steering mechanism. (*Courtesy* Massey-Ferguson, Limited)

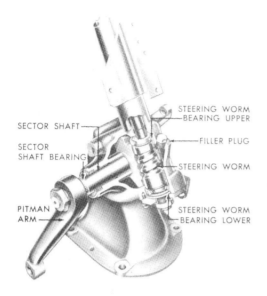

FIG. 18-6.   Worm-and-sector steering gear. (*Courtesy* Ford Motor Company)

Tractor wheel brakes are also used as an aid in steering. The brake can be applied to either wheel independently of the other wheel. This makes it possible to apply the right-wheel brake when turning right and the left-wheel brake when turning left: With this braking arrangement, it is possible to pivot three-wheel tractors on either rear wheel. Four-wheel tractors require a slightly larger turning radius.

Steering gears usually run in oil and have a reservoir separate from other tractor parts. It is necessary to check the oil and keep it at the proper level. Use the oil recommended by the manufacturer of the tractor.

Some four-wheel tractors use the double-sector and pinion system. This requires two drag links but simplifies the steering of tractors that have an adjustable front-tread width (Figure 18–7).

Power Steering

Power steering units are standard equipment on most new tractors. A vane-type hydraulic power-steering unit is shown in Figure 18–8. Oil is supplied

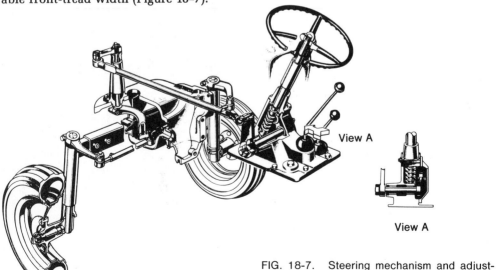

FIG. 18-7.   Steering mechanism and adjustable front end for a four-wheel tractor. (*Courtesy* Massey-Ferguson, Limited)

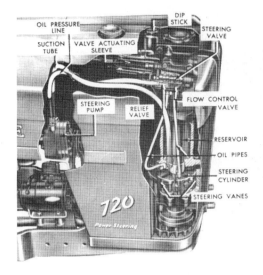

FIG. 18-8. A power-steering unit of the vane type. (*Courtesy* Deere and Company)

A cylinder-type power-steering unit is shown in Figure 18–9. It is similar to the vane-type unit in its controls and reaction to the operator turning the steering wheel.

Power-steering units ease the job of guiding a tractor by doing much of the work for the operator and by reducing the shock transmitted to the steering wheel. The operator is required to exert a maximum force of only four to five pounds on the steering wheel when a tractor is equipped with a power-steering unit. The operator still retains the feel of steering the tractor, but does not become so fatigued.

Steering Crawler Tractors

Most farm-type crawler tractors use either a clutch-and-brake arrangement for steering (Figure 18–10) or a brake-and-differential-drive arrangement (Figure 18–11). The clutch-and-brake arrangement makes it possible to

to this unit by a positive displacement pump. A valve arrangement delivers oil to either side of the steering cylinder depending upon the direction in which the operator is turning the steering wheel.

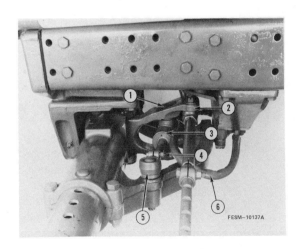

FIG. 18-9. A cylinder-type power-steering unit showing the following parts: cylinder bolt (1), center steering arm (2), power steering cylinder (3), ball joint adjusting bolt (4), ball joint assembly (5), and hydraulic line (6). (*Courtesy* International Harvester Company)

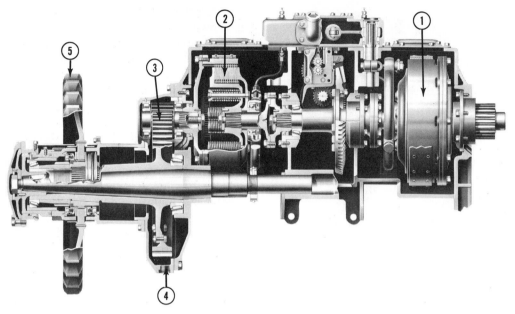

FIG. 18-10.   Clutch-and-brake steering mechanism for crawler tractor: brake drum
(1), steering clutch (2), final-drive pinion (3), final-drive gear (4), final-drive sprocket
for track (5). (*Courtesy* Caterpillar)

FIG. 18-11.   The brake-and-differential drive
for a crawler steering mechanism: drive pinion
(A), differential, spur-gear type (B), final drive
(C), steering brakes (D). (*Courtesy* White Motor
Company)

disengage either track from the driving
gears. This causes the tractor to turn in
the direction of the disengaged track because all of the power is being transmitted to the other track. Applying the
brakes on this same side causes the tractor to make a sharp turn. The tractor will
pivot on either track when one steering
clutch is disengaged and the brake is applied on the same side.

If the differential-and-brake arrangement is used, the action is similar
to that of a wheel tractor when the brake
is applied on one side. One track slows
down, while the other track actually increases its speed because of the action of
the differential. This difference in track
speeds causes the tractor to turn.

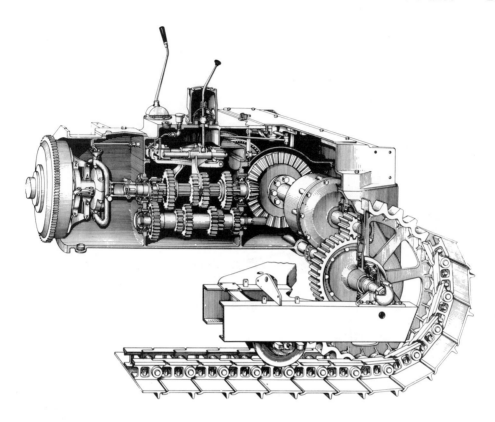

FIG. 18-12. Cutaway view of a crawler drive showing the transmission and the final-drive units. (*Courtesy* Caterpillar)

THE BRAKES

The brakes on a tractor have several functions. They are used in stopping and turning, but they are also a safety feature. Three types of brake mechanisms are in common use: the disk type, the external-band type, and the internal-expanding-shoe type.

The Disk Brake

The *disk type* of brake is installed on either the rear axle shaft or the differential shaft. One type of disk brake is shown in Figure 18–13. When the brake is applied, the friction plates are clamped between stationary plates. The friction plates are splined to the axle shaft or differential shaft and will stop or tend to stop the tractor as they are clamped between the stationary plates.

The External-Band Brake

The *external-band* type of brake has a drum attached to either the axle shaft or the differential shaft. This drum rotates with the shaft. A band with a friction surface lining fits over the drum (Figure 18–14). One end of the band is fastened to the tractor frame. When the brake is applied, a linkage between the brake pedal and the drum tightens the band on the drum. This applies a stopping force to the rear wheels of the tractor.

FIG. 18-13.   Double-disk brakes mounted on the rear axle, showing their location relative to the other parts. Actuating plates (*center*) and friction plates (*outside*) are shown in A. B shows the brakes in place on the axles. (*Courtesy* Massey-Ferguson, Limited)

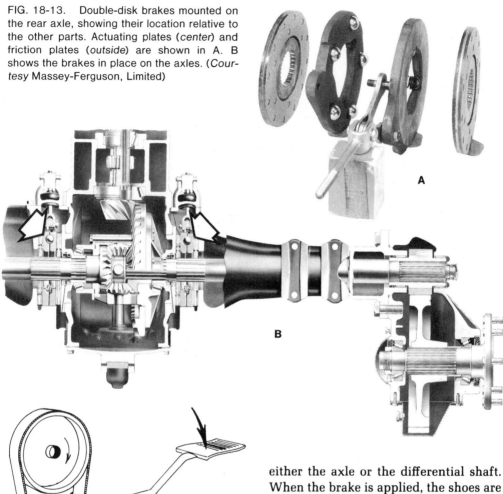

FIG. 18-14.   The external band type of brake. When the brake is applied, the band tightens on the drum. This type of brake is used on some garden tractors. (Drawing by Roger Cossette)

The Internal-Expanding Brake

The *internal-expanding* brake consists of a set of stationary brake shoes and a set of drums that are mounted on either the axle or the differential shaft. When the brake is applied, the shoes are forced against the rotating drum. This action causes a stopping force to be applied. The internal-expanding brake is also used on trucks and automobiles (Figure 18–15).

ADJUSTING BRAKES

The procedure for adjusting brakes will vary with the type of brake and the make of tractor. It is best to refer to the owner's manual for specific instructions. Most tractors have two brake pedals so that the brake can be applied on either

BRAKE SHOE   RETRACING SPRINGS

BRAKE ADJUSTING SCREW

BRAKE DRUM

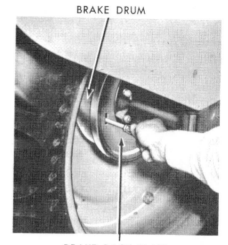

BRAKE BACK PLATE

FIG. 18-15.   Internal expanding shoe brakes. The adjusting screw is accessible through the opening in the brake drum. (*Courtesy* Ford Motor Company)

side to aid in turning. However, for road travel the brake pedals can be locked together so that they act as one pedal and apply the brakes simultaneously on both sides. It is important, then, that the brakes be adjusted so that both sides will brake the same amount when the pedals are locked together. It is best to operate the tractor and to apply the brakes to determine if both sides brake evenly. Make the necessary adjustments to get even braking action.

Belt Drives

At one time many tractors were equipped with belt pulleys that made them useful as power units to operate feed grinders, silage blowers, irrigation pumps, and many other machines. Tractors with belt pulleys are not common at present, but there are still many belt drives in use on farm machines. Most of the belt drives now in use are of the V-belt type. Some flat belt drives are also in use. The following discussion relates to flat belt drives, but the principles involved can also be applied to V-belts.

The speed at which the belt pulley turns is important because the driven machine must run at a certain speed in order to do its job effectively. The speed of the pulley is usually expressed in revolutions per minute (rpm).

The speed of the belt running over the pulley is usually expressed in feet per minute (fpm). For example, if a belt is said to be running at a speed of 3000 fpm, this means that 3000 feet of belt will pass any given point in one minute.

Since the outside surface of the pulley and the surface of the belt are in contact with each other, they must be moving at the same speed in fpm. The surface speed of the pulley can also be expressed in fpm.

The surface speed (peripheral speed) of a pulley is equal to its circumference multiplied by its speed in revolutions per minute.

$$PS = C \times rpm \qquad \text{(Equation 1)}$$

Where:

PS = peripheral speed (fpm)
C = circumference of pulley in feet*
  ($\pi \times$ diameter of pulley)
rpm = revolutions per minute

*Pulley diameter is usually given in inches, but must be changed to feet for these calculations.

These principles can be applied in several ways.

EXAMPLE 1. Let us assume that we wish to know the rpm of a certain pulley 12 inches in diameter (D) that is being driven by a belt whose speed is 3140 fpm. We can apply Equation 1 and solve it for rpm.

$$rpm = \frac{PS}{C} = \frac{3140}{\pi \times \dfrac{D}{12}}$$

$$= \frac{3140}{3.14 \times \dfrac{12}{12}} = \frac{3140}{3.14 \times 1}$$

$$= 1000$$

Notice that we changed the diameter of the pulley from inches to feet by dividing the diameter by 12. This is necessary because the peripheral speed is in feet per minute but the pulley diameter was given in inches.

EXAMPLE 2. We can develop a very useful equation for calculating pulley sizes and speeds from Equation 1. In Figure 18–17 we have a belt running over two pulleys, a driver and a driven pulley.

If we apply Equation 1, we know that both pulleys have the same peripheral speed (PS) since the same belt runs over both pulleys.

Then PS (driver) = PS (driven)

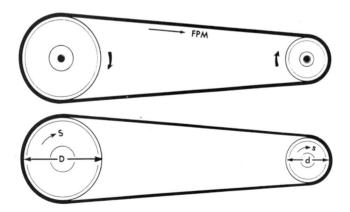

FIG. 18-16 (*Top*).   The surface or peripheral speed (PS) of two pulleys is the same when the belt runs over both pulleys. It is also the same as the belt speed. This is usually expressed in feet per minute (fpm). (Drawing by Roger Cossette)

FIG. 18-17 (*Bottom*).   The diameter (D) of the driver pulley multiplied by its speed (S) in rpm is equal to the diameter (d) of the driven pully multiplied by its speed (s) in rpm. (Drawing by Roger Cossette)

Also:

C × rpm (driver) = C × rpm (driven)

Let:

π D = circumference of driver pulley
π d = circumference of driven pulley
  S = rpm of driver pulley
  s = pm of driven pulley
PS = π D × S (driver) = π d × s
    (driven)
Then π D × S = π d × s
Divide both sides by π
Then D × S = d × s   (Equation 2)

We can say that the diameter (D) of the driver multiplied by the speed (S) of the driver is equal to the diameter (d) of the driven pulley multiplied by the speed (s) of the driven pulley. This is a very convenient equation for solving many pulley size and speed problems. It can be applied to flat belts and "V" belts.

When calculating V-belt pulley size and speed problems, Equation 2 can be used, but the "D" and "d" in the equation refer to the "pitch diameter" of the V pulley. This is somewhat less than the outside diameter of the V pulley. Pitch diameter information is available for all V pulleys.

A typical problem is as follows: An electric motor having a V pulley that has a pitch diameter of 4 inches is running at 1800 RPM. It is driving a water pump that should run at 2400 RPM. What size (pitch diameter) pulley is necessary for the water pump?

Solution: D = 4 inches
         S = 1800 RPM
         s = 2400 rpm
         d = ?

Use Equation 2, D × S = d × s and insert the known values as follows:

$$4 \times 1800 = d \times 2400$$
Solve for d
$$d = \frac{4 \times 1800}{2400} = 3 \text{ (inches)}$$

Therefore, we would select a V pulley having a pitch diameter of 3 inches.

Equation 2 can also be used when the pulley size is given in metric units. Assume that the pitch diameter of the pulley on the electric motor in the above problem is 10 cm. Determine the pitch diameter, in centimeters, of the pulley that is necessary on the water pump.

$$D \times S = d \times s$$
$$10 \times 1800 = d \times 2400$$
Solve for d
$$d = \frac{10 \times 1800}{2400} = 7.5 \text{ cm}$$

SAFETY

Belt pulleys, drive belts, and PTO shafts are always a source of danger. Loose clothing is easily caught by these power transmission units. Stay clear of moving pulleys, belts, and shafts, and use the safety shields that are furnished with the equipment.

SUMMARY

Power take-off (PTO) shafts are used to furnish rotary power to field machines and stationary units such as irrigation pumps. PTO shafts are made to conform to a standard size and speed. The original standard for most farm tractors was a 1⅜-inch six-spline shaft which ran at 540 ± 10 rpm. The new standard is a 1⅜-inch 21-spline shaft which runs at 1000 ± 25 rpm. Some tractors have PTO shafts that can be converted to either standard. Safety shields are provided for PTO shafts.

Most wheel tractors use either the worm-and-lever or the worm-and-sector system of gear reduction in the steering mechanism. Power-steering units which assist the operator are available for most tractors and are standard equipment on some models. Power-steering units use oil under pressure to assist in steering. A valve arrangement directs the oil to the proper side of the steering-unit piston or vane.

Crawler tractors have a clutch-and-brake arrangement that permits disengaging one track while power is supplied to the other track. Some crawler tractors use a brake along with a differential to steer the tractor. Applying the brake to one track causes the tractor to turn.

Tractor brakes may be one of the following types: disk, external band, or internal-expanding shoe. Brakes must be properly adjusted and maintained in order to give dependable service.

Shop Projects

A.   Determining the peripheral speed of a pulley

At the end of Chapter 9, we learned how to determine the speed of a shaft with the use of a revolution counter or tachometer. Now we will determine the peripheral speed of a pulley in feet per minute (fpm).

1.   Select a flat pulley that has an exposed shaft so that a revolution counter or tachometer can be used.

2.   Measure the diameter of the pulley in inches.

3.   Start the engine or motor to turn the pulley at any convenient speed within the range of the engine or motor.

4.   Determine the rpm of the pulley with a revolution counter or tachometer.

5.   Now determine the peripheral speed (PS) of the pulley by using Equation 1 from this chapter.

$$PS = C \times rpm$$

Remember that C is the circumference of the pulley (diameter × π) and must be in feet.

B. Getting acquainted with PTO and steering gear

Select five tractors of different makes and fill in the following chart. Most of this information can be obtained from the operator's manual for each tractor.

Do all the tractors meet SAE-ASAE standards for PTO speeds?

Tractor		PTO		Steering Gear	
Make	Model	Type	rpm	Type	How Adjusted and Lubricated
1					
2					
3					
4					
5					

Questions

1. Can two belt pulleys of different diameters have the same peripheral speed (surface speed)? Why?

2. Why should PTO shields be in place at all times when PTO-driven machines are being used?

3. Why are two clutches necessary when a tractor is equipped with a live PTO shaft?

4. What is the 1958 standard ASAE PTO-shaft speed? What advantage does it have over the previous standard?

5. What is a universal joint?

6. How does the worm-and-sector steering mechanism differ from the worm-and-lever mechanism?

7. What is the principle of most power-steering units used on farm tractors?

8. Explain how a steering mechanism for a crawler tractor differs from the steering mechanism for a wheel tractor.

9. Why should tractor brakes be kept properly adjusted?

References

Fundamentals of Machine Operation, Preventive Maintenance, Deere & Company, Moline, Illinois, 1973.

Fundamentals of Machine Operation, Tractors, Deere & Company, Moline, Illinois, 1974.

Fundamentals of Service, Power Trains, 3rd Edition, Deere & Company, Moline, Illinois, 1977.

Tractor Maintenance, American Association for Vocational Instructional Materials, Athens, Georgia, 1975.

19 Hydraulic Systems

The tractor hydraulic system is another means of making the tractor more versatile. It enables the engine to deliver power to many places on the tractor and to machines being operated by the tractor. A simple hydraulic system used on farm tractors consists of a pump, a cylinder, an oil reservoir, oil lines, and control valves. However, before discussing the hydraulic system in detail, it is necessary to explain the principles involved.

All farm tractor hydraulic systems use an oil to transmit power. Oil, like other liquids, is not compressible. It cannot be forced, by pressure, into a smaller space than it occupies when it is not under pressure. Oil, again like other liquids, takes the shape of the container that it occupies. Oil is also a lubricant and a rust inhibitor. These characteristics make it well-suited for transmitting power in the hydraulic system.

When we apply pressure to a liquid, such as oil, the pressure is transmitted to all parts of the container. Figure 19–1 shows a bottle filled with oil. Pressure is applied by forcing a cork into the filled bottle. The pressure is equal in all directions.

The simple hydraulic system in Figure 19–2 shows how a force applied to one part of the system can lift a load in another part of the system. If we increase the size of the cylinder that is lifting the load, we can lift a much heavier load with the same force. Assume that the small cylinder in Figure 19–3 has an area of one square inch and the large cylinder has an area of ten square inches. If we apply a force of one pound on the small piston, we can balance a load of ten pounds in the large cylinder because, as in the case of the bottle, the pressure is equal in all directions. The pressure in the small cylinder is one pound per square inch (area multiplied by the unit force). The pressure in the large cylinder is also one pound per square inch, but the area is ten square inches. This makes it possible to balance this ten-pound load with the one-pound force.

There is another point that we must emphasize. In Figure 19–2 it is necessary to move the piston and the force only one inch in order to raise the load one inch, because both cylinders are the same size. However, in Figure 19–3 it is necessary to move the piston and the one-pound force a distance of ten inches in order to lift the ten-pound load a distance of one inch. It is necessary to displace ten cubic inches of oil in the small cylinder in order to increase the oil volume in the large cylinder by the same amount.

In farm tractor hydraulic systems, the oil is circulated under pressure by a pump. The pump must force the oil into the cylinder. The piston in the cylinder moves and actuates a lever system. The lever system is linked to the machine to be used for the job to be done, such as raising a plow, angling a disk, or steering a tractor.

TYPES OF OIL PUMPS

The oil pump must deliver oil under pressure to the cylinder. The capacity of

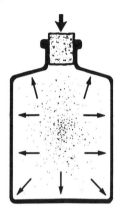

FIG. 19-1.   The pressure applied at the cork is distributed uniformly throughout the bottle. (Drawing by Roger Cossette)

the pump is determined by the pressure at which it will operate and by the quantity of oil it will pump in a given time.

The relatively high pressure at which a hydraulic system must function requires that the pumps used must be of the positive-displacement types—piston, gear, rotor, or vane.

The Piston Pump

A simple *piston pump* is illustrated in Figure 19-4. Oil is drawn into the cylinder through an intake valve on the outward stroke of the piston. As the piston moves forward, the inlet valve closes, and the oil is forced out through the outlet valve into the hydraulic system. Several pistons and cylinders of this type can be used as a unit to make up a pump for a hydraulic system. Piston pumps with several cylinders give a more uniform flow of oil and cause less vibration in the hydraulic mechanism.

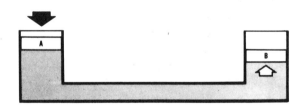

FIG. 19-2 (*Above*).   A force applied at A will balance or lift a comparable load at B. (Drawing by Roger Cossette)

FIG. 19-3 (*Below*).   A force of one pound at A will balance a load of ten pounds at B, because the area is ten square inches at B and only one square inch at A. The pressure is uniform throughout the system. (Drawing by Roger Cossette)

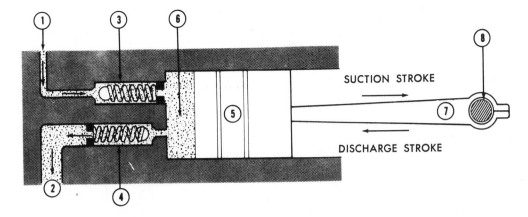

FIG. 19-4.   A single-piston-type oil pump: inlet (1), outlet (2), inlet valve (which opens on the suction stroke) (3), outlet valve (which opens on the discharge stroke) (4), piston (5), cylinder (6), connecting rod (7), and crankshaft (8). (Drawing by Roger Cossette)

FIG. 19-5.   A multipiston pump. (A) Several pistons, such as that at (1) are spaced around the central shaft. The rotating parts of this pump fit inside the stationary housing (2). (B) A more detailed view of the rotating parts. (*Courtesy* Vickers, Inc., Division of Sperry Rand Corporation)

The Gear Pump

The *gear pump* (Figure 19–6) is made with two gears, one gear that is driven and another that meshes with the first. The gears are encased in a close-fitting housing. As the gears rotate, oil is

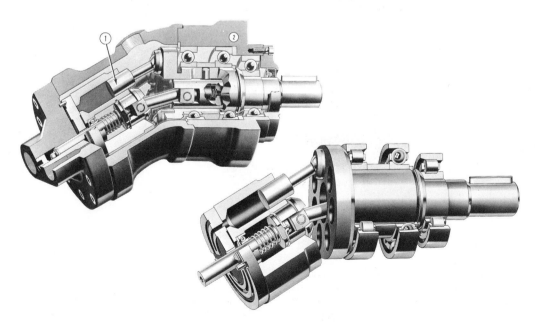

drawn in through the inlet side, carried around between the gears and the housing, and forced out through the outlet. Very little clearance is permissible between the gears and the housing. Large clearances cause oil leaks and low pump efficiency.

Gear pumps provide a uniform flow of oil to the hydraulic system. They will pump oil in either direction of rotation.

The Rotor Pump

The *rotor-type* pump (Figure 19–7) has two rotors, one inside the other. The inner rotor has external teeth that mesh with the internal teeth of the outer rotor.

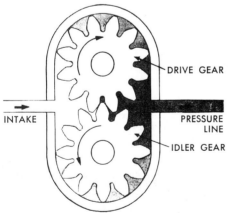

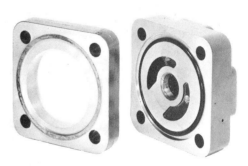

FIG. 19-6. A gear pump disassembled to show the gears that form the pumping unit. Note the close fit between the gear and the housing. (*Courtesy* J. I. Case Company)

FIG. 19-7.   A rotor pump. View of the assembled pump. (*Below*) Disassembled pump showing inner (A) and outer (B) rotors. (*Courtesy* Char-Lynn Company)

The inner rotor is mounted on a shaft that drives the pump unit. The spaces between the rotor teeth are filled with oil. Oil is drawn in on the inlet side as the pump revolves and the spaces between the rotor teeth increase in size. It is carried to the outlet side by the rotor teeth and is forced into the outlet as the spaces between the rotor teeth become smaller on this side. As in gear pumps, the clearance between the various parts must be very small so that oil cannot escape past the parts.

The Vane Pump

The *vane-type* pump is also a positive-action pump. It consists of a ring, a rotor, and vanes (Figure 19–8).

The vanes are actuated by the rotor. The rotor shaft is mounted off-center with respect to the pump ring. The vanes move in and out of slots as the rotor revolves. This causes oil to be drawn into the vane compartments on the inlet side and forced out on the outlet side. Close tolerances are necessary to avoid leakage and a loss of efficiency.

HYDRAULIC PUMP DRIVES

Most farm tractors have the hydraulic pump drive located at some point ahead of the clutch so that the hydraulic pump runs at all times when the engine is running. This is known as a live type of drive. Its advantage is that the hydraulic system will function independently of the tractor clutch.

Some tractors have the hydraulic pump drive located behind the clutch. This type of pump drive does not function when the clutch is disengaged. This may be an undesirable feature in this type of pump drive, because the hydraulic system is not independent of the clutch, as it is in the live system. It does, however, provide a positive means of disengaging the hydraulic system with the clutch pedal.

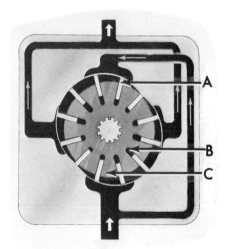 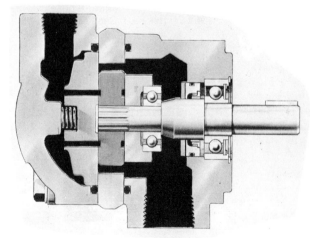

FIG. 19-8.   Cross section of a vane-type pump. The vanes (A) move in and out of the slots (B) around the rotor (C). This action causes oil to be pumped in the direction shown by the arrows. (*Courtesy* Vickers, Inc., Division of Sperry Rand Corporation)

HYDRAULIC-SYSTEM OILS

The oil that must be used in the hydraulic system varies with the make and model of tractor. The oil must flow readily at low temperatures. Crankcase oils, transmission oils, and special hydraulic-system oils are used. The oils act as a lubricant as well as a medium for transmitting power. The manufacturer's recommendations must be followed in selecting the proper oil to be used in any given system.

HYDRAULIC CYLINDERS

The hydraulic cylinder and piston are used to change the hydraulic power of the oil into mechanical power that will perform certain machine operations. A single-acting cylinder and piston are shown in Figure 19-9.

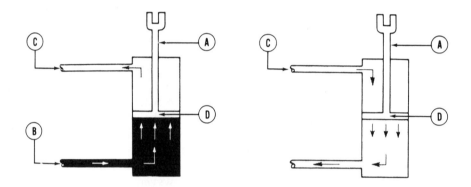

FIG. 19-9.   A single-acting (one-way) cylinder. (*Left*) The dark area indicates oil under pressure during the raising stroke. (*Right*) The weight of the implement forces the piston down when the pressure is released, causing the oil to return to the sump. Piston rod (A), oil-pressure hose (B), breather hose (no oil pressure) (C), and piston (D). (*Courtesy* J. I. Case Company)

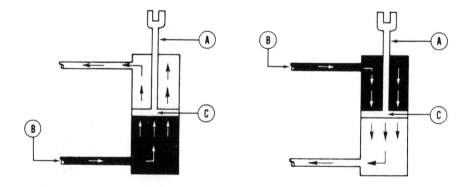

FIG. 19-10.   A double-acting (two-way) cylinder. (*Left*) The raising stroke. (*Right* The lowering stroke. The dark area shows oil under pressure on both strokes. Piston rod (A), oil-pressure hose (B), piston forced downward by hydraulic oil pressure (C). (*Courtesy* J. I. Case Company)

Oil is pumped into the cylinder for raising an implement, such as a plow, out of the ground, but the weight of the implement is utilized to force the oil out of the cylinder and back into the reservoir as the implement goes back to its working position. This type of cylinder requires only one oil line. The double-acting cylinder, illustrated in Figure 19-10, requires two oil lines and has positive action in both directions. Oil can be forced into either side of the piston. The double-acting cylinder can be used to force an implement into the ground as well as to raise the implement.

When oil is being forced into one side of the piston, the oil on the other side is permitted to flow back to the oil sump or reservoir. This is accomplished through a valve arrangement.

Pistons in hydraulic cylinders can be held in any position merely by stopping the oil flow to the piston and trapping the oil in the system with the valve mechanism.

Hydraulic cylinders can be built into the tractor to raise and lower tractor-mounted equipment (Figure 19-11) or they may be the external type (Figure 19-12) to control implements that are pulled behind the tractor.

External cylinders are controlled by the hydraulic control lever on the tractor. Oil is supplied to the cylinder through high-pressure hoses. To prevent damage the hoses are usually fitted with couplings that will disconnect from the tractor if the implement safety hitch should be released because of some obstruction.

The American Society of Agricultural Engineers has set standards for hydraulic cylinders to be used on farm implements. The implement industry has adopted these standards. The ASAE standards established various dimen-

FIG. 19-11. A hydraulic piston and cylinder (1) built into the tractor rear housing to control the mounted implements: lift arm (2), pump (3), control valve (4), control lever (5), pump-relief valve (6), cylinder safety valve (7), check valve (8), and back pressure valve (9). (*Courtesy* Ford Motor Company)

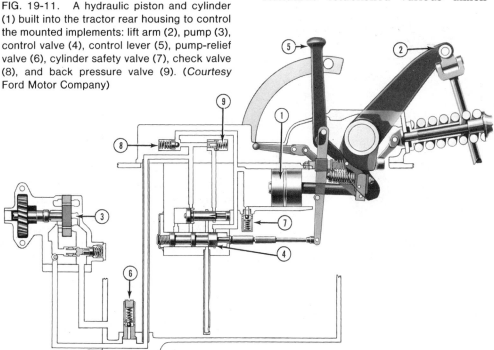

FIG. 19-12. The external or portable cylinder: (A) cylinder mounted on a plow, (B) cutaway view of a cylinder. The piston and rod are visible. (*Courtesy* J. I. Case Company and Char-Lynn Company)

sions of cylinders so that a cylinder of one make can be used on a machine of another make. The cylinder in Figure 19-12 is a standard ASAE cylinder.

HYDRAULIC-SYSTEM VALVES

A hydraulic system must have several valves to direct the flow of oil to the proper part of the system. A *relief valve* (Figure 19-11) is necessary to prevent excessive pressures. The maximum pressure developed in most systems ranges from about 1000 to 2500 pounds per square inch. When the pressure exceeds a certain maximum, the relief valve opens to permit the escape of oil back to the reservoir. This is necessary to prevent damage to the system. Hydraulic pumps are positive in action and will build up pressure to a dangerous level unless oil is permitted to escape through a relief valve.

The relief valve should function only when dangerously high pressures are reached. A weak spring may cause the valve to function when heavy, but

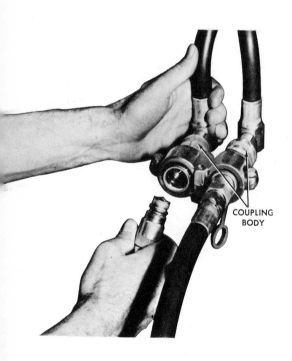

COUPLING BODY

FIG. 19-13. Hydraulic hose couplings of the breakaway type. (*Courtesy* International Harvester Company)

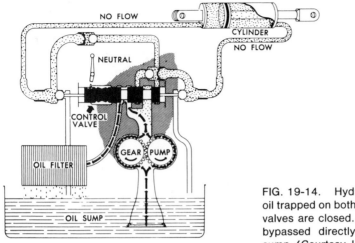

FIG. 19-14. Hydraulic system in neutral with oil trapped on both sides of the cylinder. All the valves are closed. Oil from the pump is being bypassed directly through the filter to the sump. (*Courtesy* J. I. Case Company)

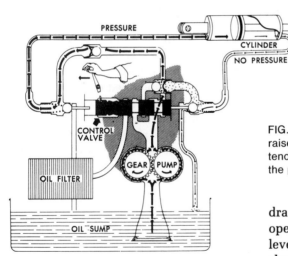

FIG. 19-15. Hydraulic system in position to raise an implement. The cylinder is being extended by the oil that is being forced into it from the pump. (*Courtesy* J. I. Case Company)

not excessive, loads are being lifted. If this takes place, the relief valve should be repaired or replaced.

Check valves (Figure 19-11) are used to permit the flow of oil in a line in only one direction.

Control valves vary from one make of hydraulic system to another, but in all systems they are used by the operator to direct the flow of oil to and from the hy-

draulic cylinder. The control valve is operated by a hand lever. Moving the lever merely slides the valve to open and shut oil passages.

Figure 19-14 shows the control valve in a complete hydraulic system with the control lever in neutral position. Oil trapped on both sides of the piston holds the piston in a stationary position. The implement being operated by the cylinder will also be held stationary. The oil pump is bypassing the oil back to the oil reservoir.

The hydraulic system in Figure 19-15 is shown in position to raise the implement. The control-valve lever has been moved forward. This shifts the valve into a position to allow oil to flow from the pump to the cylinder. The piston is forced from left to right, in the diagram, thus extending the cylinder and raising the implement. Oil to the right of the piston is forced back into the sump. When the implement is raised, the valve returns to neutral.

To lower the implement, the lever (Figure 19-16) is moved back. This places the control valve in a position to permit oil to be pumped into the right side of the cylinder and forces the piston to travel from right to left, in terms of the diagram. The implement goes to a preset working depth and stays there. The control level will automatically return to neutral when the action is complete.

OPEN-CENTER AND CLOSED-CENTER SYSTEMS

The systems shown in Figures 19-16 and 19-17 are known as open-center sys-

tems. In this system, the flow of oil remains constant at all times, but the pressure is varied. The open-center system was widely used on farm tractors until a greater demand for hydraulic power for such uses as power steering, power brakes, and additional remote cylinders became common.

The closed-center system more nearly meets the hydraulic needs of modern farm tractors. In one closed-center system an accumulator (Figure 19-18) is used to hold a reserve supply of oil under pressure.

Several types of accumulators are in use on hydraulic systems. A bladder type of accumulator is shown in Figure 19-19. When oil is pumped into the accumulator some of the space that was occupied by the bladder is now occupied by oil. The bladder may be precharged with an inert gas such as dry nitrogen. The oil that is pumped into the accumulator will be under high pressure. When a maximum preset pressure is reached, the oil from the pump is diverted back to the reservoir under low pressure. This

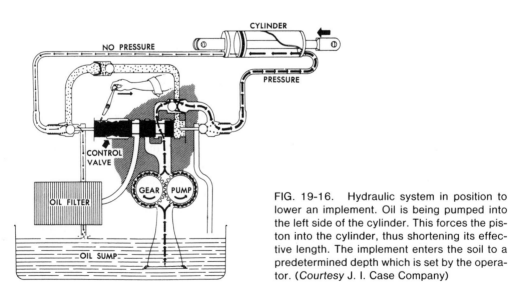

FIG. 19-16. Hydraulic system in position to lower an implement. Oil is being pumped into the left side of the cylinder. This forces the piston into the cylinder, thus shortening its effective length. The implement enters the soil to a predetermined depth which is set by the operator. (*Courtesy* J. I. Case Company)

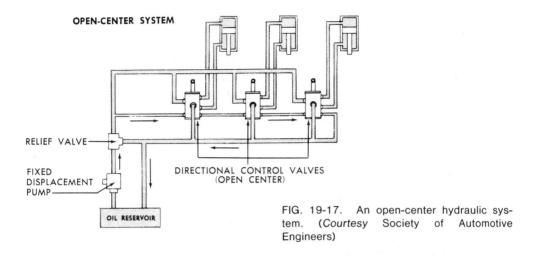

FIG. 19-17. An open-center hydraulic system. (*Courtesy* Society of Automotive Engineers)

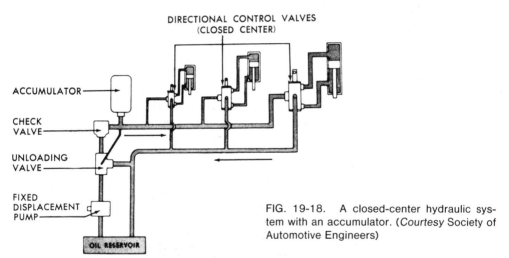

FIG. 19-18. A closed-center hydraulic system with an accumulator. (*Courtesy* Society of Automotive Engineers)

leaves more engine power available for doing other work. Also, the accumulator with oil under high pressure is available to do hydraulic work that requires a high oil flow for relatively short periods of time. When the pressure in the accumulator drops to a preset level, the pump will again supply oil to it until the pressure reaches the maximum level. The closed-center system's main advantage is that a pump with less capacity can be used. Also, the pump does not pump oil continuously as it does in the open center system. This leaves more power for other uses. The accumulator holds a reserve supply of oil under pressure that can be used in an emergency even though the engine is not running.

In another closed-center system (Figure 19-20) a variable-displacement pump is used to supply varying amounts of oil to meet the needs of the hydraulic

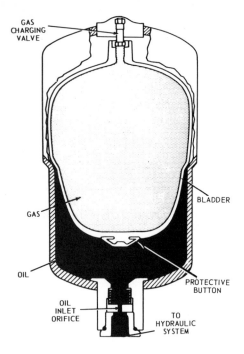

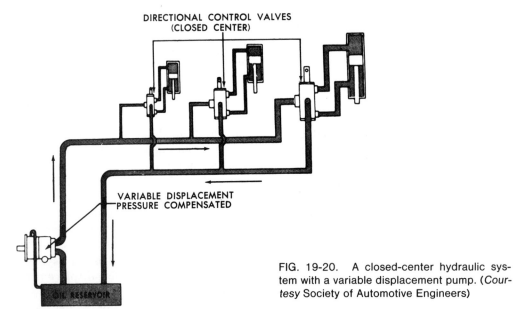

system. This pump has a high capacity to meet the maximum demands of the system and will return to a low power-consuming standby condition when there is little or no demand for oil. This system does not use an accumulator.

MAINTAINING THE HYDRAULIC SYSTEM

The hydraulic system will give many years of service if it is carefully operated and properly maintained. It should not be overloaded. However, heavy implements usually have auxiliary springs to aid in lifting. The hydraulic system is equipped with a relief valve. These are properly adjusted at the factory and should prevent overloads. If it is necessary to check the relief-valve pressure, it should be done by a competent person who has the equipment necessary to do this job.

The operator's manual should be consulted before making adjustments in

FIG. 19-19.   A bladder type accumulator. The oil in the accumulator is under pressure and ready for use when needed by the hydraulic system. (*Courtesy* Deere and Company)

FIG. 19-20.   A closed-center hydraulic system with a variable displacement pump. (*Courtesy* Society of Automotive Engineers)

the linkages and control levers of the system. It also gives instructions for doing other service jobs on the system.

Always maintain the proper level of hydraulic fluid in the system. This is necessary to get maximum piston travel in the cylinder and to provide lubrication to the parts that are lubricated by the hydraulic fluid. The hydraulic fluid must be kept clean in storage and in the system. Always wipe dirt off the filler plug before removing it. Use clean containers and the type of oil recommended by the manufacturer. Keep the hose connections clean.

The oil in the system should be changed once a year or according to the manufacturer's instructions. Some systems are equipped with magnetic drain plugs. This type of plug attracts metal particles and keeps them from circulating in the system. It is important to clean such a plug before replacing it. If the hydraulic system is equipped with an oil filter, it should be serviced at the same time that the fluid is changed.

SUMMARY

The hydraulic system is a great laborsaver because it extends the use of the tractor. It turns the tractor into a versatile tool by making it possible to control mounted and pull-type equipment hydraulically. The simple movement of a lever is all that is necessary to raise or lower an implement, to lift a heavy load onto a truck, or to do many other difficult jobs.

The basic principles are simple. Oil cannot be compressed, but it can be pumped under pressure to remote cylinders as well as to built-in cylinders. Hydraulic control valves permit the operator to direct the flow of oil to and from the hydraulic cylinder.

The types of pumps used are the piston, gear, rotor, and vane types.

Hydraulic-system oils vary with the type of system. Crankcase oils, transmission oils, and special hydraulic-system oils are used. In the hydraulic system, oil is also a lubricant and a rust inhibitor.

Hydraulic systems must be properly maintained in order to give many years of trouble-free service. Use clean oil of the proper type. Keep the oil at the proper level to insure maximum piston travel. Service the hydraulic system according to the manufacturer's recommendations.

Shop Projects

A. Studying the pump

1. Obtain one pump of each type—piston, gear, rotor, and vane. (These need not be pumps from tractor hydraulic systems, but can be old weed-sprayer or water pumps. We are interested in the principle and construction of each pump.)

2. Disassemble the pumps one at a time and note the working parts that are necessary in each case.

3. Notice the valves that are present on each pump. Determine why these valves are necessary.

4. Sketch the principal working parts of each pump and explain how each one pumps fluids.

5. Assemble the pumps. Make certain that the gaskets are in good condition and properly installed. New gaskets may be necessary if the pump is to be put into service.

6. Complete a report on these pumps and hand it in to your instructor.

B. Changing oil in the hydraulic system

Hydraulic systems vary from one make of tractor to another. For this reason, it is important to obtain an operator's manual for the tractor selected for the oil-changing exercise. Follow the manual's instructions.

General procedure

1. Clean dirt from the external parts of the filler plug and the drain plug to prevent dirt from getting into the system.

2. Drain the oil according to the instructions in the operator's manual.

3. If the hydraulic system has an oil filter, it should be cleaned or re-placed. (See the operator's manual.)

4. Clean the drain plug, if it is of the magnetic type. Make sure all metal particles are removed before replacing the plug.

5. Replace the drain plug and fill the reservoir with the correct amount of oil of the recommended type.

6. Replace the filler plug.

Note: It must be emphasized that the foregoing instructions are general. Consult the operator's manual for your tractor's exact instructions.

Questions

1. Why is oil well-suited for transmitting power in a hydraulic system?

2. What characteristics do all types of hydraulic-system oil pumps have in common?

3. Piston pumps for tractor hydraulic systems are usually made with several small pistons rather than one large piston. Why?

4. What is the advantage of a live type of hydraulic-pump drive?

5. What types of oil are commonly used in hydraulic systems? What determines the type of oil that must be used in a given system?

6. What is a single-acting hydraulic cylinder? A double-acting hydraulic cylinder?

7. Why is a relief valve used in most hydraulic systems?

8. What is a check valve? Why is it used?

9. What causes wear in the piston and cylinder? What should be done to keep this at a minimum?

10. What is the ASAE standard hydraulic cylinder? Why was it developed?

11. How often should the oil be changed in the hydraulic system?

12. What is a magnetic drain plug? Why is it used?

13. Explain the difference between an open-center and a closed-center hydraulic system.

14. Why is it important that hydraulic fittings be kept clean?

References

Farm Equipment Hydraulics, Implement and Tractor Publications, Inc., Kansas City, Missouri, 1966.

Fundamentals of Service, Hydraulics, 2nd Edition, Deere & Company, Moline, Illinois, 1972.

Hydraulics, Care & Operation, Volume 1, American Association for Vocational Instructional Materials, Athens, Georgia, 1974.

Hydraulics, Inspecting and Testing, Volume 2, American Association for Vocational Instructional Materials, Athens, Georgia, 1974.

Tractor Maintenance, American Association for Vocational Instructional Materials, Athens, Georgia, 1975.

20 Safe Tractor Operation

Many farm accidents involve a tractor or a tractor operated machine. Tractor accidents causing injury or death are a serious matter of widespread concern. It is estimated that in one year there are nearly 50,000 accidents in which the tractor is involved. Recently approximately 800 fatal tractor accidents occurred in one year. This large number of accidents results because either the operator fails to understand when the tractor is in a dangerous situation or he is careless.

The National Safety Council (Figure 20-1) made a study of 317 fatal accidents involving tractors. This study showed that over 30 percent of the accident victims were under 20 years of age, and many were under five years of age. Therefore, it is obvious that an alarmingly large percentage of youngsters are involved in tractor accidents. Neither children nor older people should be permitted to ride on tractors as extra passengers.

A Purdue University study shows that the time required for a tractor to get into a tipping position is very short. In tipping backward, a tractor will reach the point of instability in as short a time as 1.36 seconds. A tractor operator has very little time to act once his tractor is approaching a dangerous position.

CENTER OF GRAVITY

A tractor is not dangerous when properly operated. However, it is easy to get a tractor into an unsafe position when the operator is not careful. We must study the tractor in order to understand why it will tip over under some conditions. A tractor, like all other objects, has a center of gravity, a point where we can consider the entire weight to be concentrated. A yardstick has its center of gravity at the 18-inch mark, halfway between the ends. The yardstick will balance at this point. (Figure 20-2.)

The location of a tractor's center of gravity will vary with the make and model. On a wheel tractor, it is at a point ahead of the rear axle and midway between the rear wheels (Figure 20-3). The entire weight of the tractor can be considered to be concentrated at this point. As long as the vehicle remains level enough so that a vertical line drawn through the center of gravity falls within the points where the wheels make contact with the ground, the tractor will not tip (Figure 20-4). However, when the tractor is on a side slope such as that illustrated in Figure 20-5 and the vertical line through the center of gravity falls outside of the points where the wheels

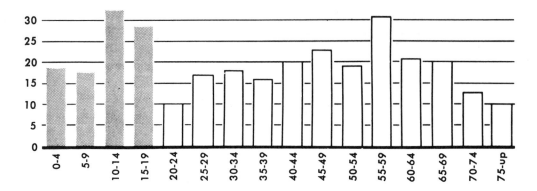

FIG. 20-1.   Over 30 percent of the victims of this study of 317 fatal accidents involving tractors were under 20 years of age. (*Courtesy* National Safety Council)

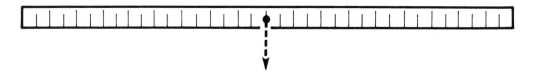

FIG. 20-2.   A yardstick has its center of gravity at its center. (Drawing by Roger Cossette)

FIG. 20-3.   The center of gravity is ahead of the rear axle and halfway between the rear wheels on a wheel tractor. (Drawing by Roger Cossette)

make contact with the ground, the tractor will tip over.

There are many instances of using a tractor when the center of gravity is outside the points where the wheels contact the ground. Driving on a side slope, running one wheel in a ditch, hitting an ob-struction with the uphill wheel, or dropping into a hole with the downhill wheel are some of the more common situations that can be dangerous and, in many cases, cause a tractor to tip over.

When the tractor is making a turn, there is an additional force acting on it.

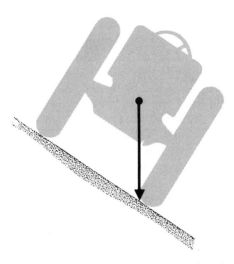

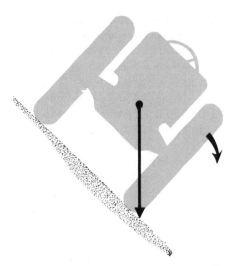

FIG. 20-4. A tractor will not tip to the side as long as a vertical line through the center of gravity falls within the points where the rear wheels contact the ground.

FIG. 20-5. When a vertical line through the center of gravity falls outside the points where the rear wheels contact the ground, the tractor will tip. (Drawings by Roger Cossette)

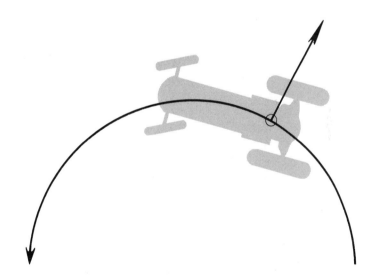

FIG. 20-6. Turning at high speed causes a centrifugal force outward. This force can cause the tractor to tip. High-speed turns are dangerous. (Drawing by Roger Cossette)

Centrifugal force acts outward and is added to other forces that may cause the tractor to tip sideward (Figure 20-6).

Making a sharp turn on a hillside is especially dangerous when the turn is made so that the centrifugal force acts downhill. Sharp turns at high speed are extremely dangerous under all conditions. When a tractor traveling at high speed strikes an object, such as a rock or stump, the force is sometimes sufficient to tip the tractor. Also, when one drive wheel drops into a hole, the effect will be about the same as hitting an object. It is always wise to drive slowly if you are not familiar with field conditions.

The tractor will tip backward when the front wheels are raised to a point at which the vertical line through the center of gravity falls behind the line through the points where the rear wheels make contact with the ground (Figure 20-7). This can be caused by excessive load at the drawbar, especially with a high hitch or when going up a steep hill. The front end of a tractor may be raised dangerously high when driving up a steep grade or through a ditch. Pulling a load at the drawbar under these conditions adds to the chance that the

tractor will tip backward. When it is necessary to go up a steep grade, it is safest to back the tractor up the grade with the load attached to the front.

A tractor will tend to become lighter in the front when a load is being pulled at the drawbar. It may actually rear up, to the point where the tractor will tip. This is especially true when the load is hitched above the drawbar at the rear axle, or on the upper linkage of a mounted hitch system. Always avoid a high hitch.

The fact that a high hitch will tend to cause a tractor to tip backward can be demonstrated by substituting a small table for the tractor. Tie a rope to two legs about 16 inches (drawbar height) from the floor (Figure 20-8) and notice how many pounds of horizontal pull are necessary to raise the other two legs of the table on the side opposite from the rope. Now move the rope up to a point about 24 inches above the floor and notice how much less force is necessary to raise the opposite two legs of the table. Always pull a load by hitching it to the drawbar. Never hitch a load to the rear

FIG. 20-7.   A tractor will also tip backward when a vertical line through the center of gravity falls behind the point where the rear wheels contact the ground. (Drawing by Roger Cossette)

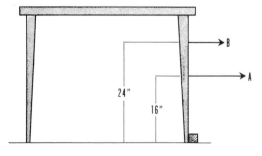

FIG. 20-8.   It takes less force to tip the table when pulling at point (B) than at point (A), because the hitch is higher. The same principle applies to a high tractor hitch. Always hitch to the drawbar. It is safer. (Drawing by Roger Cossette)

axle or to the upper linkage of a mounted hitch system.

SAFETY ON THE HIGHWAY

In conducting everyday farming operations, it is often necessary to drive a tractor on a road or highway where high-speed traffic is usual. This always presents a hazard to the tractor operator as well as to the motorist. It is difficult for the motorist to judge the slow speed of the tractor, since he is accustomed to faster-moving traffic. An accident can occur if the motorist misjudges the relative speeds of the two vehicles. If such an accident takes place, the tractor operator finds himself in a very vulnerable position on the tractor and can receive serious injuries, even in what might otherwise be a minor accident.

The tractor operator has a big responsibility when he takes slow-moving equipment on the highways. He owes it to himself, to his family, and to motorists to make his vehicle visible from a long distance. He should keep his vehicle as far to the right as possible, to give the motorist a chance to pass.

Several types of safety-warning devices can be used. One of the most effective warning devices is the slow-moving vehicle emblem that was developed as the result of a study in Ohio (Figure 20-9). This triangular-shaped emblem was designed for mounting on the rear of slow-moving vehicles. The emblem's reflective red border and fluorescent orange center make it visible for at least one-sixth of a mile. This gives the motorist time to decide what action is necessary to avoid an accident. The emblem is not a substitute for lights or other safety devices required by law. It has the endorsement of the National Safety Council and the American Society of Agricultural Engineers. This emblem is generally recognized as the official slow-moving vehicle (SMV) emblem. Several states have laws requiring its use when slow-moving vehicles are operated on highways. The size and shape of the SMV emblem have been standardized. It is available from several commercial sources.

State laws usually require specific emblems, flags, reflectors, or lights at various times of day. Those transporting farm machinery on any road should become familiar with the laws governing the movement of machinery.

The National Safety Council emphasizes the following points in making the operation of tractors on highways as safe as possible.

1. Good planning can help you avoid much unnecessary movement of farm equipment on heavily traveled roads.

FIG. 20-9.   A tractor equipped with (1) a slow-moving-vehicle (SMV) emblem and (2) a roll bar. The roll bar will protect the operator if the tractor should tip. (Photo by Authors)

2. Keep the tractor under control. Slow down for turns. Leave the tractor in gear on downgrades.

3. Only experienced operators, possessing mature judgment, should be given the responsibility of handling farm machinery on public roads. Permit no extra riders on the tractor.

4. Well-located entrances to fields and farmyards allow good visibility for operators of farm machinery and motor vehicles.

5. Courtesy makes friends and prevents accidents. Pulling off the pavement to let faster traffic pass takes only minutes—and may save lives.

6. Proper use of emblems, red flags, and approved lighting calls the attention of the motorist to the dangers of slow-moving farm machinery.

7. Know and conform to local traffic laws.

OTHER SAFETY PRACTICES

Always be careful when refueling a tractor. Shut off the engine before putting in fuel. Do not smoke or have an open flame nearby while refueling. Fumes from gasoline and other fuels, including LP gas, are heavier than air and tend to gather in low places. Refuel tractors outside buildings. This will usually prevent the accumulation of fuel fumes and will make the job of refueling much safer.

Always keep power take-off shields in place when using PTO-driven equipment. PTO shields have been standard-

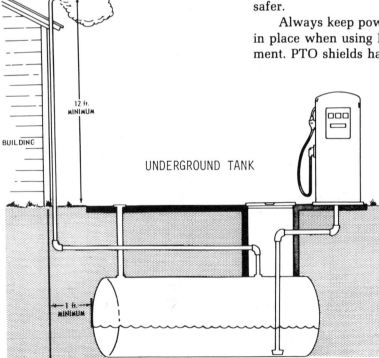

FIG. 20-10. Underground fuel storage system. (*Courtesy* Deere and Company)

FIG. 20-11. PTO safety shields should always be in place when using PTO-driven equipment. This safety shield is the nonremovable type (A). The underside of the shield is shown in (B). (*Courtesy* National Safety Council)

ized so that shields from one make of equipment will fit on another. Avoid being exposed to turning shafts, moving belts, chains, pulleys, gears, and other moving parts.

Always stop the machine before lubricating, adjusting, or cleaning it. Wear close-fitting clothing when working with power machinery.

Tractor coolants and hydraulic fluids can be dangerous. Be careful not to remove the radiator cap when the coolant is hot. Coolants are under pressure. Hydraulic fluids can be dangerous because of heat and high pressures of 2,000 psi or more.

Tractor noise, as well as noise from other sources, can cause operator fatigue and nervousness if he must be near the source of the noise. Most new farm tractors are designed so that the noise level of the tractor is below dangerous levels. This is especially true of tractors that have cabs designed to keep tractor noise from being readily transmitted to the cab. However, many older tractors do not have cabs and some do not use a muffler. These conditions may cause the noise at the operator's platform to be at a level which is detrimental to the operator. Hearing losses can result from continued exposure to high levels of noise.

Sound intensity is expressed in decibels. Sound intensities that can be barely heard have a decibel reading of 0. Sound that is so loud that it is on the threshold of causing pain has a decibel reading of 120 to 130. These and other levels of sound are shown in Table 20-1.

The loudness or intensity of sound or noise is doubled with every 10 decibel increase in intensity. In other words, a 70 decibel sound is twice as loud as a 60 decibel sound. Sound intensities at a decibel level of 80 to 85 are generally considered to be the maximum safe level for long time exposure. The sound intensity at the operator's platform of most new farm tractors is at or near the 80 to 85 decibel level.

OSHA Regulations

The Occupational Safety and Health Act, enacted by Congress in 1970, has broad implications for industry and agriculture. Occupational safety and health standards for agriculture have been developed. Particular emphasis is placed upon power take-off shielding, proper instructions of employee operators, and standards for noise levels. Since these standards are subject to change, it is important that those operating farm tractors, and particularly those employing drivers, keep informed of these standards.

SUMMARY

Tractors will tip sideward or backward in less than a second and a half under certain conditions. This gives the operator a minimum of time to correct the situation.

The center of gravity of a wheel tractor is at a point ahead of the rear

TABLE 20-1.   SOUND LEVELS*

Sound	Intensity Level Decibels
Noise on the threshold of pain	120–130
Air Hammer	100–115
Farm Tractor with no muffler	100–110
Farm Tractor with muffler	80–95
Heavy street traffic	70–80
Ordinary conversation	60
Whisper	20
Rustling leaves	10
Barely audible sound	0

* Adapted from information in the following books: *Tractors and Their Power Units* by E. L. Barger et al., John Wiley & Sons, Inc., New York. 1963. *Principles of Physics* by F. Bueche, McGraw-Hill Book Co., New York, 1965.

axle and halfway between the drive wheels. A vertical line through the center of gravity must fall inside an area bounded by the points where the wheels contact the ground or the tractor will upset. Centrifugal force due to turning can cause a tractor to tip sideward.

All drawbar loads should be hitched to the drawbar and not to the axle or other points above the drawbar. A high hitch is dangerous because it causes the front end of the tractor to "rear up" with the result that the tractor may tip backward.

Safety practices must always be kept in mind when working around the farm. Machines and their moving parts are a potential safety hazard, especially in the hands of a careless operator. The tractor operator must learn to recognize unsafe practices, as well as to know when a tractor or machine is approaching an unsafe situation. He or she must understand the limitations of the machines being operated.

PTO shafts and other moving parts, the battery, and the engine fuel are also potentially dangerous. Tractor coolants and hydraulic fluids are potentially dangerous because of high temperatures. Hydraulic fluids are maintained under very high pressures. Always be careful when working with these and other possible hazards.

Noise or sound can be detrimental to the operators of farm equipment. Safe noise levels are at or below the 80 to 85 decibel level.

When operating the tractor on a highway, always take measures to make certain that the motorist is properly warned of the fact that a slow-moving vehicle is on the road. The slow-moving vehicle emblem, red flags, reflectors, and lights can be used to help warn motorists.

Shop Projects

A.   Demonstrating the dangers of a high hitch

1.   Select a model toy tractor and add some weight to its rear wheels. A few ounces of washers or similar weight attached to each rear wheel should serve the purpose. Lock the wheels by placing a rod through the spokes from one side to the other.

2.   Provide a small wire hook at rear axle height and also attach a small wire hook to the drawbar.

3.   Select one large or two small rubber bands.

4. Place the toy tractor on a level surface. (It may be necessary to fasten a sheet of sandpaper to the surface to keep the wheels from sliding.) Hook the rubber bands to the tractor at axle height (Figure 20-12). Pull straight back on the rubber bands and measure the distance the bands are stretched before the front end of the tractor is raised from the table.

5. Now hitch the same rubber bands at drawbar height and measure how much farther the bands have to be stretched before the front end of the tractor is raised. Also notice that a greater pull is required.

FIG. 20-12.   A hitch at axle height (1) will cause a tractor to rear up with a relatively light load and it may tip backward. In this case, a rubber band (2) and a toy tractor are used to demonstrate this principle. (Photo by Authors)

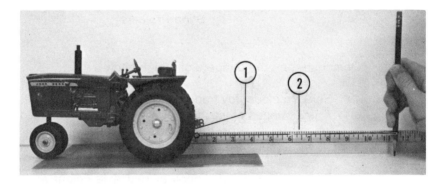

FIG. 20-13.   The same toy tractor with the load at drawbar height (1), rubber bands (2) hitched to the drawbar. A greater pull is required to raise the front wheels off the ground. It is safer to hitch to the drawbar. (Photo by Authors)

6.   This will demonstrate that a high hitch is dangerous. Always hitch to the drawbar. It is safer.

Note: A fisherman's scale can be substituted for the rubber bands. The force can then be measured in each case.

B.   Demonstrating the dangers of rotating PTO shafts and other moving parts

The following equipment is necessary:

1 smooth rod, ¼" × 4", that has been scraped lengthwise with a file
1 portable electric drill and bench holder or vise to hold the drill
1 piece of metal tubing that will fit loosely over the rod
Cloth stripped into pieces 1" wide and 8-10" long

Proceed as follows with the demonstration (see Figure 20-14):

1.   Place the rod in the jaws of the drill chuck. Turn the drill on.

2.   Take a strip of cloth and hold the end of it against the moving rod. The rod represents the PTO shaft. The rod will catch the cloth and wind it around itself.

3.   Turn the drill off and place the tube over the rod. Turn the drill on again. As you hold the tube in place, have someone dangle the cloth over the tube. In this case, there is no chance for the cloth to get caught on the rod with this protective shield over it.

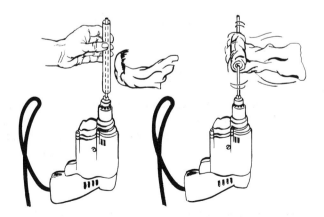

FIG. 20-14.   Demonstrating the dangers of an exposed rotating shaft. (*Courtesy National Safety Council*)

We can conclude from this demonstration that even a round shaft will catch clothing. Shields will protect the operator. Always keep the safety shields in place. (This demonstration was developed by the National Safety Council.)

C.   Prepare a display or bulletin board of tractor accident statistics or news articles.

Questions

1.   Why is the rate of fatal tractor accidents so high among people under 20 years of age?

2.   What can be done to reduce the number of tractor and machinery accidents?

3.   Why does a tractor tip backward more easily when the load is hitched above the drawbar than when it is hitched to the drawbar?

4.   About how much time is required to get a tractor into a tipping position when it is improperly hitched or loaded?

5.   Will a PTO safety shield from one make of machine fit on another make? Why?

6.   Why can rotating shafts, pulleys, and gears be dangerous?

7.   How does centrifugal force influence a tractor when making a turn?

8.   How can we make highway travel with tractors and machinery more safe?

9.   Why should the tractor engine be shut off when refueling?

10. What is the National Safety Council? How is this organization helping with farm safety problems? (To find answers to these questions, ask your instructor or the school librarian for literature on the National Safety Council.)

11. What is the SMV emblem?

12. Is the SMV emblem required by law in your state?

References

Barger, E. L., *et al., Tractors and Their Power Units,* John Wiley and Sons, Inc., New York, New York, 1963.

Fundamentals of Machine Operation, Agricultural Machinery Safety, Deere & Company, Moline, Illinois, 1974.

Fundamentals of Machine Operation, Tractors, Deere & Company, Moline, Illinois, 1974.

Leaflets, National Safety Council, Farm Division, Chicago, Illinois.

21 Selection and Management of Tractors and Machines

Mechanization of farming operations has greatly reduced the number of agricultural workers required to produce food and fiber for our population.

One farm worker can do much more work in the same length of time with a machine and tractor than he could by hand or with horse-drawn equipment. In the United States, far more than enough food and fiber for the entire population is produced by less than one-tenth of the population. Efficient machines of high capacity help make this possible. These machines are carefully engineered to give excellent performance.

However, the farmer must make a substantial investment to own them. He must select, manage, and maintain them with great care in order to make his investment a sound one. The selection of the proper equipment for a given farm is a problem that can best be solved by having a knowledge of what various machines can accomplish when properly operated.

HOW TO SELECT MACHINES

Which make or model of machine to select can be decided by considering the following questions:

1. Is the machine made by a reliable manufacturer whose dealer maintains a complete parts-and-service department in your community?
2. Is the machine adapted to your needs? Consider the size, attachments, and simplicity of design.
3. Is it the best buy? Consider price, quality, operating costs, and adaptability to your farming operation.
4. On engine-driven machines, does the engine use a fuel that is readily available?
5. Are adjustments and minor repairs easily made?
6. Is the machine easily serviced?
7. Does the machine provide factors that are to the advantage of the operator? Consider quietness, ease of starting, and protection against exhaust gases and moving parts.
8. Is the machine well-designed and neat in appearance?

When the type of machine necessary to do the job has been determined, decide the size to be purchased. The number of days in the season available to use this machine and the size of tractor available to operate it should be determining factors. To help make a decision on the size of field machine necessary, it is useful to know how many acres can be worked with it in one hour. The following formula will be helpful to you.

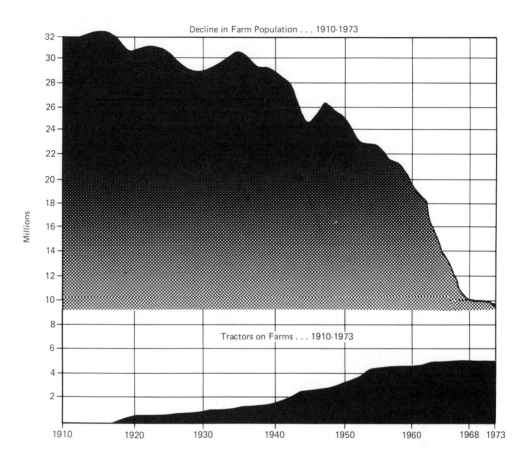

FIG. 21-1.   The farm population continues to decline as mechanization increases. The total number of tractors on farms (as shown above) has started to decline, but the total horsepower of those tractors is increasing as larger tractors are being manufactured and sold. (Adapted from U.S. Department of Agriculture Statistics)

$$C = \frac{WS}{10}$$

C = Capacity of machine in acres per hour

W = Working width of machine in feet

S = Speed of machine in miles per hour

EXAMPLE:

How many acres per hour can be seeded with a 12-foot grain drill traveling at 5 mph?

SOLUTION:

$$C = \frac{WS}{10} = \frac{12 \times 5}{10} = 6 \text{ acres per hour}$$

This formula allows for about 17 percent of time lost in turning, filling the grain box, and other servicing jobs. It is then possible to determine the number of hours or days that will be necessary to perform a given field operation. In the foregoing example, it would take ten hours to seed a 60-acre field or 40 hours to seed 240 acres. We usually assume that only half of the working season is available for field work. Wet weather and other factors usually account for the other half. In other words, if the actual seeding time for a crop requires 10 days, then 20 calendar days should be allowed to get this job done.

The size of the tractor that should be purchased is determined by the size of the machine to be operated. In the case of tillage machines, size is also influenced by the type and condition of the soil. It is especially difficult to determine the size of the tractor necessary for a tillage machine because of the great variation in power necessary to suit varying soil conditions. The following table shows the variation in draft of many field machines.

Because of the large variation in implement draft, it is not wise to recommend a general formula for determining the tractor size necessary for a given machine. It is best to reply on the experi-ences of local farmers and implement men, if local draft tests are not available.

COST OF OWNING EQUIPMENT

Modern farming practices demand a large investment in machinery. Machinery is necessary to reduce labor costs. Most farm jobs can be done most economically when performed by a machine. However, it is necessary to do some calculating to determine whether or not a given machine is a good investment. To be worthwhile, the investment in the machine has to reduce the cost of performing farm operations and increase the net return to the farmer. If this cannot be accomplished, the investment in the machine is not advisable.

The cost of owning farm machinery is affected by fixed and operating costs.

Fixed costs are those that add up whether or not the machine is used.

Fixed costs are:

1. Initial investment
2. Depreciation
3. Interest on the investment
4. Taxes, insurance, housing, etc.

TABLE 21-1.   DRAFT OF FIELD MACHINES

Machine	Draft Pounds per Foot of Width	
Moldboard plow	300 to 900	
One-way disc tiller	160 to 350	
Single disc	40 to 130	
Tandem disc	81 to 160	
Spike-tooth harrow	20 to 60	
Field cultivator with sweep shovels	57 to 200	
Grain drill	30 to 80	
Corn planter	80 to 100	(pounds per row)
Mower	60 to 100	

Operating costs are those that add up only when the equipment is in use.

Operating costs are:

1. Fuel, oil, and lubrication
2. Repairs
3. Labor to operate the machine

In addition to these, the cost of owning a machine is also affected by the number of hours per year that the machine is used. More annual use reduces the hourly cost of owning a machine. Each of the above items requires further discussion.

FIXED COSTS

Initial Investment

The *initial investment* is the amount that the machine cost when it was purchased, whether new or used. It is the amount of money that must be invested in order to own the machine.

Depreciation

Depreciation, an unavoidable loss, is the amount that the machine decreases in value each year through use and obsolescence. There are several methods of calculating depreciation. One simple method is to subtract the junk value of a machine from its purchase price and divide this amount by the expected life of the machine. For example, if we assume that a machine cost $2,000.00 when new, has a junk value of $100.00, and an expected life of ten years, its annual depreciation is:

$$\frac{\$2,000 - 100}{10} = \$190.00$$

The depreciation at any time in the life of a machine can be determined by subtracting its resale value from its original cost.

Interest on the Investment

Interest on the investment is one of the costs of ownership, because the money that is invested in the machine could be used for purchasing land, bonds, or other income-producing enterprises were it not invested in the machine. A convenient way to determine the amount to charge to interest is to base it on the average investment in the machine. This results in equal annual charges. The average investment is one-half the initial cost plus the junk value or

$$\frac{\text{initial cost} + \text{junk value}}{2}$$

If we take the previous example of a machine that cost $2,000.00 and had a junk value of $100.00, the average investment would be

$$\frac{\$2,000 + \$100}{2} = \$1,050.00$$

At an interest rate of 9 percent, the annual charge for the investment would be $1,050.00 × .09 or $94.50.

Taxes

Taxes are assessed against farm equipment in much the same manner as other farm property is taxed. This tax is a part of the cost of owning the machine. An annual tax charge of 1 percent of the original cost of the machine is commonly used for convenience in calculating. The annual tax charge on a $2,000.00 machine would be $2,000.00 × .01, or $20.00.

Insurance

Insurance to cover losses by fire, wind, and other hazards also must be charged against the cost of owning a machine. The farmer who does not carry insurance takes the risk himself. This justifies the charge in all cases. The annual charge for insurance will amount to about .25 percent of the original cost of the machine.

Housing for Machinery

Housing for machinery is not always provided, but its cost should be charged against the cost of machinery ownership. When housing is not provided, the more rapid rate of depreciation of some machines will probably be equal to the cost of housing. Providing housing for machinery does not always produce a return on the investment, although it is conducive to good management practices. Machines that are housed can be repaired more easily during stormy weather. A corner of the machine shed can be converted to a shop. This provides a convenient place to do repair work. The cost of housing machinery will vary with the amount invested in the building. It is advisable to use a low-cost but durable structure as a machine shed. When this is done, the annual cost of housing is about 1 percent of the new cost of the machine. On a $2,000.00 machine, the annual cost of housing will be $2,000.00 × .01, or $20.00.

OPERATING COSTS

Engine Fuel

Engine fuel is one of the major items of cost in operating engine-driven machines. Studies made in Kansas show that a two-plow tractor will use about 16 gallons of fuel per working day. In the same study, it was found that a four-plow tractor used about 31 gallons of fuel per day. Fuel consumption can vary with carburetor adjustment and load on the engine, but these figures may be considered average.

Engine Crankcase Oil

Engine crankcase oil consumption, according to the Kansas study, varied from .6 gallons per day for a two-plow tractor to 1.1 gallons per day for a four-plow tractor. The oil consumption will also vary with the mechanical condition of the tractor. A tractor that is in good condition will use less oil per day than

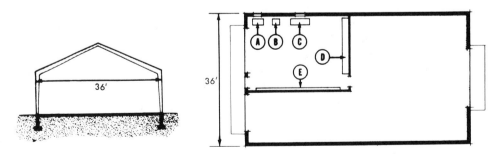

FIG. 21-2.   Machine shed floor plan with the shop shown in one corner: welder (A), grinder (B), power tools (C), workbench (D), bins for bolts and other parts (E). (Drawing by Roger Cossette)

one that has worn engine parts such as pistons, rings, or cylinder walls.

The above figures on oil consumption include an oil change at approximately every 60 hours of tractor operation. This makes the figures seem high according to present rates of oil consumption. Improved engines and oils have permitted change periods on many tractors to be extended to about 250 hours. This would reduce the above oil-consumption figures.

Lubricating Greases and Oils

Lubricating greases and oils (excluding crankcase oil, which has already been discussed) must also be considered as a part of the operating costs. These lubricants include pressure-gun, wheel-bearing, transmission, and differential greases, as well as oils used for lubricating many of the simple bearings found on farm machinery. The Kansas study shows that the annual cost of these greases and oils amounts to about .4 percent of the original cost of the machine.

The cost of greases and oils is relatively small and less than the cost of the labor necessary to do the lubrication job. This emphasizes the importance of using good greases and oils, because the labor required is the same when inferior lubricants are used.

Machinery costs must be analyzed carefully so that the owner will know whether or not a machine is a good investment. Sometimes it may be cheaper to hire than to own a machine to do a certain job. In other cases, it may be more economical to run one machine for longer hours than it is to own two sets of equipment to get a certain job done.

Machinery costs become less per hour when the machine is used a greater number of hours per year. (See Figure 21-3).

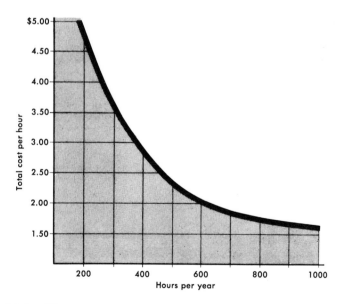

FIG. 21-3.   The cost of a machine per hour of use becomes smaller as the machine is used a greater number of hours per year. (*Courtesy* North Dakota State University)

Labor

The *cost of labor* required to operate a machine must be considered in calculating machine costs. At least one person is required to operate most machines; therefore, that person is not available to do other productive work. Prevailing wages should be used in making these cost calculations.

Labor cost calculations are especially important when machinery is used for doing custom work. The machine owner quite often furnishes labor for this type of service. He must then figure labor costs.

Nebraska Tractor Tests

In Chapter 4 we discussed power, horsepower, and other terms that apply to engine size and performance. The horsepower that a tractor will develop is of interest to the buyer. However, an unbiased source for this information is desirable because comparisons between makes and models of tractors are then based on the same test procedures.

Nebraska Tractor Tests are a reliable source of information on farm tractors and are a helpful management tool. These tests were established, by law, in Nebraska in 1919 and have been in use since that time. The law requires that a stock model of every tractor sold in Nebraska must be tested and passed by a board of three engineers under State University management.

Almost every manufacturer of farm tractors wants to sell tractors in Nebraska, so a stock model of nearly all farm tractors on the market has been tested at Nebraska. These tests are made public and are available from the University of Nebraska to anyone interested in having the information. The test results are a reliable means of comparing tractors.

A typical test report is shown below. Notice that the results are shown in metric units as well as English units.

NEBRASKA TRACTOR TEST 1267 — JOHN DEERE 4040 DIESEL

POWER TAKE-OFF PERFORMANCE

Power Hp (kW)	Crank shaft speed rpm	gal/hr (l/h)	lb/hp.hr (kg/kW.h)	Hp.hr/gal (kW.h/l)	Cooling medium	Air wet bulb	Air dry bulb	Barometer inch Hg (kPa)
		Fuel Consumption			Temperature °F (°C)			

MAXIMUM POWER AND FUEL CONSUMPTION

Rated Engine Speed—Two Hours (PTO Speed—1002 rpm)

| 90.80 | 2200 | 6.732 | 0.516 | 13.49 | 192 | 51 | 75 | 28.980 |
| (67.71) | | (25.483) | (0.314) | (2.657) | (88.8) | (10.4) | (23.9) | (97.861) |

VARYING POWER AND FUEL CONSUMPTION—Two Hours

79.90	2280	6.038	0.526	13.23	189	52	76	......
(59.58)		(22.856)	(0.320)	(2.607)	(87.2)	(10.8)	(24.2)	
0.00	2364	2.455	.....	.....	182	51	75	......
(0.00)		(9.292)			(83.1)	(10.6)	(23.9)	
40.89	2327	4.130	0.704	9.90	186	52	76	......
(30.49)		(15.634)	(0.428)	(1.950)	(85.3)	(10.8)	(24.2)	
91.30	2200	6.800	0.519	13.43	191	52	75	......
(68.08)		(25.741)	(0.316)	(2.645)	(88.3)	(10.8)	(23.9)	
20.66	2351	3.295	1.111	6.27	184	51	75	......
(15.40)		(12.471)	(0.676)	(1.235)	(84.4)	(10.6)	(23.9)	
60.44	2302	4.961	0.572	12.18	186	52	76	......
(45.07)		(18.780)	(0.348)	(2.400)	(85.6)	(11.1)	(24.4)	
Av 48.86	2304	4.613	0.658	10.59	186	51	75	29.000
Av (36.44)		(17.463)	(0.400)	(2.087)	(85.6)	(10.8)	(24.1)	(97.929)

DRAWBAR PERFORMANCE

Power Hp (kW)	Drawbar pull lbs (kN)	Speed mph (km/h)	Crank-shaft speed rpm	Slip %	gal/hr (l/h)	lb/hp.hr (kg/kW.h)	Hp.hr/gal (kW.h/l)	Cool-ing med	Air wet bulb	Air dry bulb	Barom. inch Hg (kPa)
					Fuel Consumption			Temp. °F (°C)			

Maximum Available Power—Two Hours 7th (B-2) Gear

| 74.83 | 5301 | 5.29 | 2199 | 6.43 | 6.657 | 0.620 | 11.24 | 188 | 50 | 53 | 28.410 |
| (55.80) | (23.58) | (8.52) | | | (25.201) | (0.377) | (2.214) | (86.4) | (9.7) | (11.7) | (95.936) |

75% of Pull at Maximum Power—Ten Hours 7th (B-2) Gear

| 61.70 | 4098 | 5.65 | 2306 | 4.77 | 5.627 | 0.635 | 10.96 | 185 | 33 | 39 | 28.894 |
| (46.01) | (18.23) | (9.09) | | | (21.302) | (0.386) | (2.160) | (85.1) | (0.4) | (4.1) | (97.571) |

50% of Pull at Maximum Power—Two Hours 7th (B-2) Gear

| 42.62 | 2751 | 5.81 | 2330 | 3.03 | 4.630 | 0.757 | 9.21 | 180 | 32 | 36 | 28.915 |
| (31.78) | (12.24) | (9.35) | | | (17.525) | (0.460) | (1.814) | (82.2) | (0.0) | (1.9) | (97.642) |

50% of Pull at Reduced Engine Speed—Two Hours 12th (B-4) Gear

| 42.22 | 2729 | 5.80 | 1409 | 3.11 | 3.302 | 0.545 | 12.79 | 176 | 35 | 40 | 28.930 |
| (31.48) | (12.14) | (9.34) | | | (12.498) | (0.331) | (2.519) | (80.0) | (1.7) | (4.2) | (97.692) |

MAXIMUM POWER IN SELECTED GEARS

67.57	8936	2.84	2290	14.88	3rd (A-3) Gear			178	29	29	29.310
(50.39)	(39.75)	(4.56)						(81.1)	(-1.6)	(-1.6)	(98.975)
74.28	6848	4.07	2200	8.89	5th (B-1) Gear			187	47	52	28.470
(55.39)	(30.46)	(6.55)						(86.1)	(8.3)	(11.1)	(96.139)
76.06	5804	4.91	2200	6.95	6th (C-1) Gear			188	46	51	28.480
(56.72)	(25.82)	(7.91)						(86.4)	(7.8)	(10.6)	(96.173)
77.02	5460	5.29	2199	6.54	7th (B-2) Gear			187	45	50	28.520
(57.43)	(24.29)	(8.51)						(86.1)	(7.2)	(10.0)	(96.308)
77.85	4591	6.36	2201	5.35	8th (C-2) Gear			187	47	52	28.480
(58.05)	(20.42)	(10.23)						(86.1)	(8.3)	(11.1)	(96.071)
76.81	4106	7.02	2200	4.85	9th (B-3) Gear			187	48	52	28.430
(57.28)	(18.26)	(11.29)						(86.1)	(8.9)	(11.1)	(96.004)

LUGGING ABILITY IN RATED GEAR 7th (B-2)

Crankshaft Speed rpm	2199	1975	1766	1538	1314	1096
Pull—lbs	5460	5994	6141	6180	6260	6162
(kN)	(24.29)	(26.66)	(27.32)	(27.49)	(27.85)	(27.41)
Increase in Pull %	0	10	12	13	15	13
Power—Hp	77.02	75.30	68.82	60.28	52.07	42.80
(kW)	(57.43)	(56.15)	(51.32)	(44.95)	(38.83)	(31.92)
Speed—Mph	5.29	4.71	4.20	3.66	3.12	2.60
(km/h)	(8.51)	(7.58)	(6.76)	(5.89)	(5.02)	(4.19)
Slip %	6.54	7.42	7.69	7.56	7.83	7.56

Department of Agricultural Engineering

Dates of Test: November 12 to 28, 1977

Manufacturer: JOHN DEERE WATERLOO TRACTOR WORKS, P.O. Box 270, Waterloo, Iowa 50704

FUEL, OIL AND TIME: Fuel No. 2 Diesel **Cetane** No. 50.8 (rating taken from oil company's typical inspection data) **Specific gravity converted to 60°/60° *(15°/15°)*** 0.8366 **Fuel weight** 6.966 lbs/gal *(0.837 kg/l)* **Oil** SAE 30 **API service classification** CD, CC and SD **To motor** 3.769 gal *(14.267 l)* **Drained from motor** 3.463 gal *(13.109 l)* **Transmission and final drive lubricant** John Deere Hy-Gard Transmission and Hydraulic Oil **Total time engine was operated** 50 hours

ENGINE Make John Deere Diesel **Type** 6 cylinder vertical **Serial No.** 6404DR-25 560262RG **Crankshaft lengthwise Rated rpm** 2200 **Bore and stroke** 4.25″ × 4.75″ *(108.0 mm × 120.6 mm)* **Compression ratio** 16.2 to 1 **Displacement** 404 cu in *(6625 ml)* **Cranking system** 12 volt **Lubrication** pressure **Air cleaner** paper primary and safety elements with dust evacuator **Oil filter** one screw-on cartridge **Oil cooler** engine coolant heat exchanger for crankcase oil, radiator for transmission and hydraulic oil **Fuel filter** one snap-on cartridge **Muffler** vertical **Cooling medium temperature control** two thermostats.

CHASSIS: Type standard **Serial No.** 4040H 001113R **Tread width rear** 60″ *(1524 mm)* to 118.3″ *(3004 mm)* **front** 51.8″ *(1314 mm)* to 71.8″ *(1824 mm)* **Wheel base** 104″ *(2642 mm)* **Center of gravity** (without operator or ballast, with minimum tread, with fuel tank filled and tractor serviced for operation) Horizontal distance forward from center-line of rear wheels 32.1″ *(815 mm)* Vertical distance above roadway 38.2″ *(971 mm)* Horizontal distance from center of rear wheel tread 0.06″ *(1 mm)* to the left **Hydraulic control system** direct engine drive **Transmission** selective gear fixed ratio with partial (2) range operator controlled power shift **Advertised speeds mph *(km/h)*** first 2.0 *(3.1)* second 2.5 *(4.0)* third 3.2 *(5.2)* fourth 4.1 *(6.6)* fifth 4.5 *(7.2)* sixth 5.3 *(8.6)* seventh 5.7 *(9.2)* eighth 6.8 *(10.9)* ninth 7.4 *(11.9)* tenth 8.2 *(13.1)* eleventh 8.8 *(14.1)* twelfth 9.4 *(15.1)* thirteenth 10.4 *(16.7)* fourteenth 11.2 *(17.9)* fifteenth 13.5 *(21.7)* sixteenth 17.1 *(27.5)* reverse 3.1 *(5.0)*, 4.0 *(6.4)*, 7.2 *(11.6)*, 8.5 *(13.7)*, 9.1 *(14.7)*, 10.8 *(17.4)* **Clutch** wet multiple disc hydraulically power actuated and operated by foot pedal **Brakes** wet disc hydraulically power actuated and operated by two foot pedals which can be locked together **Steering** hydrostatic **Turning radius** (on concrete surface with brake applied) right 142.9″ *(3.63 m)* left 142.9″ *(3.63 m)* (on concrete surface without brake) right 158.5″ *(4.03 m)* left 158.5″ *(4.03 m)* **Turning space diameter** (on concrete surface with brake applied) right 295.8″ *(7.51 m)* left 295.8″ *(7.51 m)* (on concrete surface without brake) right 326.9″ *(8.30 m)* left 326.9″ *(8.30 m)* **Power take-off** 1002 rpm at 2200 engine rpm, 540 rpm at 2200 engine rpm.

TRACTOR SOUND LEVEL WITH CAB	dB(A)
Maximum Available Power—Two Hours	79.5
75% of Pull at Maximum Power—Ten Hours	79.0
50% of Pull at Maximum Power—Two Hours	79.0
50% of Pull at Reduced Engine Speed—Two Hours	76.5
Bystander in 16th (D-4) gear	92.0

TIRES, BALLAST AND WEIGHT		With Ballast	Without Ballast
Rear Tires	—No., size, ply & psi *(kPa)*	Two 18.4-34; 8; 18 *(120)*	Two 18.4-34; 8; 18 *(120)*
Ballast	—Liquid (each)	382 lb *(173 kg)*	None
	—Cast Iron (each)	140 lb *(64 kg)*	None
Front Tires	—No., size, ply & psi *(kPa)*	Two 10.00-16; 6; 32 *(220)*	Two 10.00-16; 6; 32 *(220)*
Ballast	—Liquid (each)	None	None
	—Cast Iron (each)	60 lb *(27 kg)*	None
Height of drawbar		18.5 in *(470 mm)*	18.5 in *(470 mm)*
Static weight with operator—rear		8755 lb *(3971 kg)*	7710 lb *(3498 kg)*
front		3300 lb *(1497 kg)*	3180 lb *(1443 kg)*
total		12055 lb *(5468 kg)*	10890 lb *(4941 kg)*

REPAIRS and ADJUSTMENTS: No repairs or adjustments.

REMARKS: All test results were determined from observed data obtained in accordance with SAE and ASAE test code or official Nebraska test procedure. Temperature at injection pump return was 155°F *(68.5°C)*. Six gears were chosen between 15% slip and 15 mph *(24.1 km/h)*.

We, the undersigned, certify that this is a true and correct report of official Tractor Test **1267.**

LOUIS I. LEVITICUS
Engineer-in Charge

G. W. STEINBRUEGGE, Chairman
W. E. SPLINTER
K. VON BARGEN
Board of Tractor Test Engineers

EXPLANATION OF TEST REPORT

GENERAL CONDITIONS

Each tractor is a production model equipped for common usage. Power consuming accessories may be disconnected only when the means for disconnecting can be reached from the operator station. Additional weight can be added as ballast if the manufacturer regularly supplies it for sale. The static tire loads and the inflation pressures must conform to recommendations in the Tire Standards published by the Society of Automotive Engineers.

PREPARATION FOR PERFORMANCE RUNS

The engine crankcase is drained and refilled with a measured amount of new oil conforming to specifications in the operators manual. The fuel used and the maintenance operations must also conform to the published information delivered with the tractor. The tractor is then limbered-up for 12 hours on drawbar work in accordance with the manufacturer's published recommendations. The manufacturer's representative is present to make appropriate decisions regarding mechanical adjustments.

The tractor is equipped with approximately the amount of added ballast that is used during maximum drawbar tests. ·ior to the maximum power run the tire tread-bar height must be at least 65% of new tread height.

POWER TAKE-OFF PERFORMANCE

Maximum Power and Fuel Consumption. The manufacturer's representative makes carburetor, fuel pump, ignition and governor control settings which remain unchanged throughout all subsequent runs. The governor and the manually operated governor control lever is set to provide the high-idle speed specified by the manufacturer for maximum power. Maximum power is measured by connecting the power take-off to a dynamometer. The dynamometer load is then gradually increased until the engine is operating at the rated speed specified by the manufacturer for maximum power. The corresponding fuel consumption is measured.

Varying Power and Fuel Consumption. Six different horsepower levels are used to show corresponding fuel consumption rates and how the governor causes the engine to react to the following changes in dynamometer load: 85% of the dynamometer torque at maximum power; minimum dynamometer torque, ½ of the 85% torque; maximum power, ¼ and ¾ of the 85% torque. Since a tractor is generally subjected to varying loads the average of the results in this test serve well for predicting the fuel consumption of a tractor in general use.

DRAWBAR PERFORMANCE

All engine adjustments are the same as those used in the belt or power take-off tests.

Varying Power and Fuel Consumption With Ballast. The varying power runs are made to show the effects of speed-control devices (engine, governor, automatic transmission, etc.) on horsepower, speed and fuel consumption. These runs are made around the entire test course which has two 180 degree turns with a minimum radius of 50 feet. The drawbar pull is set at 4 different runs as follows: (1) as near to the pull at maximum power as possible and still have the tractor maintain the travel speed at maximum horsepower on the straight sections of the test course; (2) 75% of the pull at maximum power; (3) 50% of the pull at maximum power; and (4) maintaining the same load and travel speed as in (3) by shifting to a higher gear and reducing the engine rpm.

Maximum Power with Ballast. Maximum power is measured on straight level sections of the test course. Data are shown for not more than 6 different gears or travel speeds. Some gears or travel speeds may be omitted because of high slippage of the traction members or because the travel speed may exceed the safe limit for the test course. The manufacturer's representative has the option of selecting one gear or speed over eight miles per hour. The maximum safe speed for the Nebraska Test Course has been set at 15 mph. The slip limits have been set at 15% and 7% for pneumatic tires and steel tracks or lugs, respectively. Higher slippage gives widely varying results.

Varying Drawbar Pull and Travel Speed with Ballast. Travel speeds corresponding to drawbar pulls beyond the maximum power range are obtained to show the "lugging ability" of the tractor. The run starts with the pull at maximum power; then additional drawbar pull is applied to cause decreasing speeds. The run is ended by one of three conditions: (1) maximum pull is obtained, (2) the maximum slippage limit is reached, or (3) some other operating limit is reached.

SOUND MEASUREMENT

Sound is recorded during each of the Varying Power and Fuel Consumption runs as the tractor travels on a straight section of the test course. The dB(A) sound level is obtained with the microphone located near the right ear of the operator. Bystander sound readings are taken with the microphone placed 25 feet from the line of travel of the tractor.

An increase of 10 dB(A) will approximately double the loudness to the human ear.

246

MACHINERY STORAGE

Machinery management includes preparing machines for storage and providing proper storage facilities.

The end of each season of machine use is a good time to clean the machines, tighten all bolts, make needed adjustments, and remove all old oil and grease. Machines should be thoroughly lubricated at this time because the bearings will then be filled with lubricant for the entire storage season. Following this procedure will insure that dirt and water will be kept out of the bearings.

Machines should be thoroughly checked for needed replacement parts before the storage season. A list of these parts should be prepared. The parts can then be purchased or ordered for repairs to be made during the storage season.

Parts such as moldboards on plows and cultivator shovels should be protected with rustproofing compounds. This will make it much easier to start these machines functioning properly at the beginning of the next season.

Organize the space in the machine shed so that it will be used efficiently and so that machines may be removed in the order in which they will be needed. A driveway through the machine shed is desirable for ease in removing machines, although it can be an inefficient use of space unless it is used to store self-propelled equipment such as trucks, tractors, and other machines that can be easily moved.

If it is necessary to leave machines outside during the winter months, besides observing the general precautions previously mentioned, the machines should be lined up in an open field in straight rows at right angles to prevailing winds so that snow will drift past them. Raise them off the ground by putting pieces of wood under wheels, disks, and shovels. Belts, canvases, and other parts that can be easily damaged by weather should be removed from machines and stored in a weatherproof building.

SHOP EQUIPMENT

The kind and amount of shop equipment that is necessary on the farm depends on what kind and how much work is done.

A good way to build up equipment is to repair and clean all the tools that you already have and build storage racks for them. Add tools as they are needed and as finances permit. Always buy good tools.

Prepare a list of tools that are wanted. These can be divided into three main headings: hand tools, power equipment, and special tools.

Hand tools include saws, hammers, planes, squares, pliers, wrenches, screwdrivers, chisels, punches, files, and many other small tools such as a spark-plug gauge, a feeler gauge, and other engine-servicing tools.

The *power tools* most commonly used in farm shops include grinders, portable electric drills, power saws, a welder, and an air compressor.

Special tools are those that are needed for jobs that are not commonly done on the farm.

There are many tools that an individual farmer may want to consider. A large farm that is highly mechanized may have a complete repair shop, including engine-overhauling equipment, metal-turning lathes, and other tools used by repair shops. However, most

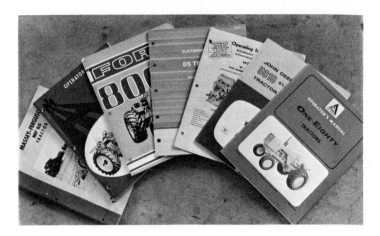

FIG. 21-4.   Owner's or operator's manuals should be kept on all tractors and machines. Keep them in a safe place. (Photo by Authors)

farms will have only those tools necessary to do the general maintenance and minor repair work on the farm. It is advisable to let dealers' service shops do the overhaul jobs on engines and machinery. These shops have the necessary specialized tools and trained mechanics to do this type of work.

SUMMARY

The job of selecting and managing farm equipment is one that requires an understanding of the capacity of various machines and tractors, and also a knowledge of the factors that add to the cost of owning equipment.

The capacity of a machine can be calculated approximately by the use of the formula $C = \dfrac{WS}{10}$. However, the power required to operate a field machine is not so easily determined, because of the wide variations that exist in soil conditions. Local authorities should be consulted for advice on this matter.

The cost of owning equipment can be divided into fixed and operating costs. Fixed costs include initial investment, depreciation, interest on investment, taxes, insurance, and housing for machinery. Operating costs on equipment include fuel, oil, greases, repairs, and labor.

An understanding of the above factors makes it possible to decide whether or not a particular machine will pay for itself.

Good maintenance practices are necessary to help keep machinery costs down. Machines must be properly lubricated, repaired, and protected from adverse weather conditions. It costs less to do a good job maintaining a machine than to pay for expensive repairs caused by lack of proper maintenance. A well-equipped shop will save much time during rush seasons and will permit the repairing of equipment when time is available.

Shop Projects

A. Using the formula $C = \dfrac{WS}{10}$ described earlier in this chapter, determine the acres per hour that can be covered with several of the field machines found on your home farm. How do the results compare with your actual experience in using the machines?

B. Refer to Table 21–1. Determine the horsepower required to pull 36 feet of tandem disc at five miles per hour in heavy soil (under high draft conditions). See Chapter 4 for the horsepower formula.

Questions

1. What points should be considered when deciding what make and model of machine to buy? Which of these points is most important?

2. How should the size of the machine be determined?

3. What might be the economic results of owning a tractor that is too large or too small for your farm?

4. What factors make up the cost of owning farm equipment? What costs go on whether or not the machine is used? What costs are present only when the machine is used?

5. Does it pay to provide housing for farm machinery? Explain.

6. Why is it important to do a good job of maintaining farm equipment?

7. When machinery is housed, what system should be used in placing the machines in storage?

8. What can be done to help protect machines that are not housed?

References

Agricultural Engineering Yearbook, American Society of Agricultural Engineers, St. Joseph, Michigan, 1976.

Fenton, F. C. and G. E. Fairbanks, *The Cost of Using Farm Machinery,* Bulletin #74, Kansas State University, Manhattan, Kansas, 1954.

Fundamentals of Machine Operation, Machinery Management, Deere & Company, Moline, Illinois, 1975.

Fundamentals of Machine Operation, Preventive Maintenance, Deere & Company, Moline, Illinois, 1973.

Planning Machinery Protection, American Association for Vocational Instructional Materials, Athens, Georgia, 1974.

Shop Planning, American Association for Vocational Instructional Materials, Athens, Georgia, 1975.

Tractor Maintenance, American Association for Vocational Instructional Materials, Athens, Georgia, 1975.

Appendix A
Metric System

The units of the metric system that are fundamental to the system are the meter (the unit of length) and the kilogram (the unit of mass).

Some of the metric units are shown in the table below:

TABLE OF METRIC MEASURES

Length				**Area**		
Unit	**Symbol**	**Value in Meters**		**Unit**	**Symbol**	**Value in Sq. Meters**
Millimeters	mm	0.001		Sq. Millimeters	mm^2	0.000001
Centimeters	cm	0.01		Sq. Centimeters	cm^2	0.0001
Decimeters	dm	0.1		Sq. Decimeter	dm^2	0.01
Meter	m	1.0		Sq. Meter	m^2	1.0

Volume				**Cubic Measure**		
Unit	**Symbol**	**Value in Liters**		**Unit**	**Symbol**	**Value in Cu. Meters**
Milliliters	ml	0.001		Cu. Millimeters	mm^3	0.000000001
Liter	l	1.0		Cu. Centimeters	cm^3	0.000001
				Cu. Meter	m^3	1.0

Mass		
Unit	**Symbol**	**Value in Grams**
Milligram	mg	0.001
Centigram	cg	0.01
Gram	g	1.0
Kilogram	kg	1000.00

Appendix B
Conversion Factors

Distance

1 inch = 2.54 centimeters (cm)
1 foot = 30.48 cm
1 mile = 1.61 kilometers (km)

1 cm = 0.39 inches
1 km = 1000 cm = 0.62 mile

Weight

1 pound = 454 grams (gm)
1 pound = 0.454 kilograms (kg)
1 kilogram (kg) = 2.20 pounds

Volume

1 cubic inch = 16.39 cubic centimeters (cc)
1 cubic foot = 28.32 liters (1)

1 liter = 1000 milliliters = 1000 cubic centimeters = 61.02 cubic inches

Power

1 HP = .746 kilowatts
1 kilowatt = 1.34 H.P.

Temperature

The following equation can be used to convert degrees Fahrenheit (F) to degrees Celsius (C)
$C = 5/9 \ (F - 32)$

Appendix C
Special Project on Troubleshooting Engines

TROUBLESHOOTING

There are many things that can go wrong with farm equipment, especially engine-driven machines. These are often minor, and the operator who knows his machines has little difficulty in locating them. Trouble should be remedied promptly.

At this point in our study of engines we have become familiar with ignition systems, fuel systems, valves, air intake systems, and engine operating principles. This is a good time to become more familiar with engines by performing a troubleshooting exercise.

Troubleshooting consists of analyzing the problem in a systematic manner. For example, if an engine fails to start or stops suddenly, the best procedure is to make a systematic check of the entire ignition system, from the battery to the distributor and spark plugs, including all adjustments. This should be followed by a complete check of the fuel system from the fuel supply to the carburetor.

The air intake system must also be checked to make sure that there is a free flow of air from the intake stack through the air cleaner, the carburetor, and the intake manifold.

The adjustment of the valves should also be checked if the engine fails to start after the ignition and fuel systems are functioning properly.

Troubleshooting on other machines should be done in a similar systematic and thorough manner. Farm equipment companies do an excellent job of providing owner's and operator's manuals for the various machines they manufacture. These manuals give the essential instructions necessary for properly operating and maintaining the equipment. They are an excellent aid to any troubleshooting job on a machine. Owner's and operator's manuals are available from dealers and branch houses.

Use the following as a guide to the causes of engine troubles.*

A Guide to Engine Troubleshooting

The following guide may be helpful to you in analyzing the causes of poor engine performance.

1. Engine fails to start:

    No fuel in tank, clogged fuel line,

* Adapted from *Trouble Shooting Guide*—Gulf Oil Company

253

or fuel pump not functioning
Fuel supply valve closed
Carburetor flooded or float stuck
Water in fuel supply
Defective or wet spark plugs
Defective fuel injector (diesel engines)
Ignition system out of time or defective (broken wires, defective switch, loose battery cable, short circuits, etc.)

2. Engine runs roughly, misfires, and backfires:

    Carburetor out of adjustment
    Water in fuel
    Leaking cylinder-head gasket
    Leaking intake-manifold gasket
    Dirty spark plugs
    Valves or tappets stuck
    Improper timing of ignition system or fuel injectors
    Loose or nearly broken ignition wire
    Moisture in ignition parts
    Cold engine
    Clogged fuel injector in diesel engine

3. Engine overheats:

    Lack of water or oil
    Load too heavy
    Clogged radiator or cooling system
    Slipping fan belt
    Ignition out of time
    Carburetor improperly adjusted
    Excess carbon in cylinders

4. Engine knocks:

    Excess carbon in cylinders
    Wrong fuel or lean fuel mixture
    Loose bearings on connecting rods, piston pins, crankshaft, or camshaft

Lack of oil
Lack of water
Broken piston rings or loose pistons
Sticking valve stems or valves improperly adjusted
Improper timing of ignition
Overload

5. Lack of compression pressure:

    Sticking valves or valves improperly adjusted
    Stuck, worn, or broken piston rings
    Worn pistons and cylinders
    Leaking cylinder-head gaskets

6. Lack of oil pressure:

    Lack of oil in crankcase
    Oil screen or oil line plugged
    Defective pressure indicator
    Severely worn bearings
    Defective oil pump
    Leaking or broken oil line

7. Lack of fuel at carburetor:

    Clogged fuel line, screen, strainer, or valve
    Vent hole in fuel tank filler cap plugged
    Fuel low in tank
    Broken fuel line or loose connection

8. Explosion in exhaust pipe:

    Late ignition
    Weak spark
    Defective exhaust valve

9. Smoky exhaust:

    Black smoke
        Overloaded diesel engine
        Fuel-injector timing too late

Rich air-fuel mixture
Poor grade of fuel
Blue smoke
 Worn pistons, cylinders, and rings
 Engine misfiring

10. Excess fuel consumption:

 Choke on
 Clogged air cleaner or intake cap
 Improper grade of oil in air cleaner
 Leaking connections on fuel lines
 Carburetor out of adjustment (too rich)

Old, worn spark plugs
Late timing

11. Defective or weak spark:

 Spark plugs defective or of wrong type
 Spark plugs wet or improperly adjusted
 Defective coil or condenser
 Breaker points worn or out of adjustment
 Magneto out of time with engine
 Impulse coupling not functioning properly

Shop Projects

A.   Gasoline engine troubleshooting *
Proceed on the assigned engine according to the following outline. (Complete each step before going to the next one.)

 1.   Check the ignition system.

  (a)   Start at the battery and check the entire wiring system; make sure that all connections are tight and properly made.

  (b)   Check all adjustments (breaker-point gap, spark-plug gap, spark timing, etc.) and correct as necessary.

  (c)   Check the entire ignition system for short circuits and other defects. Make the necessary repairs.

  (d)   Test the ignition system. Turn the ignition switch on and crank the engine. Check the firing at each spark plug or at the wire leading to the plug by leaving a short gap between the wire and the "ground."

 2.   Check the fuel system and the carburetor.

  (a)   Check the fuel supply.

* The instructor should put the engine out of order; the ignition system and valves should all be put out of adjustment or otherwise made inoperable.

(b)   Make certain that the line from the supply tank to the carburetor is free from obstructions. Blow air through the line, if necessary.

(c)   Check the carburetor for obstructions. Remove any dirt that may be clogging the jets or the fuel supply to the carburetor. Clean the carburetor screen, if necessary.

(d)   Make the necessary adjustments on the load and the idling jets.

3.   Check the air supply system. Make sure that there is a free flow of air from the air intake stack through the entire system. A clogged air cleaner or air supply line can keep an engine from running properly.

4.   Make sure that the valves are properly adjusted and in time. (Valves will not be out of time unless some serious trouble has developed. However, they may be so far out of adjustment that the engine will not run properly if at all.)

(a)   Adjust the valve-stem clearance (tappet clearance) on all valves. See the operator's manual for the correct clearance.

(b)   Check the valve timing.

5.   Start the engine.

6.   Have the instructor check your work.

Questions

1.   In doing a troubleshooting exercise on an engine, where is it best to start? Explain.

2.   How will a dirty air cleaner affect the fuel consumption on a gasoline engine? Why?

3.   Why do engines start losing power when the air cleaner is dirty or clogged?

4.   The valve clearance on an engine is much too great. How does this influence the opening and closing of the valves? Will it be early or late?

Index